Values and Virtues in the Military

Studies in Military Psychology and Pedagogy

Edited by Hubert Annen

Vol. 14

Nadine Eggimann Zanetti

Values and Virtues
in the Military

PETER LANG

Bibliographic Information published by the Deutsche Nationalbibliothek
The Deutsche Nationalbibliothek lists this publication in the Deutsche Nationalbibliografie; detailed bibliographic data is available online at http://dnb.d-nb.de.

Library of Congress Cataloging-in-Publication Data
A CIP catalog record for this book has been applied for at the Library of Congress.

This work was accepted as a PhD thesis by the Faculty of Arts and Social Sciences, University of Zurich in the fall semester 2017 on the recommendation of Prof. Dr. Willibald Ruch and Prof. Dr. Bruno Staffelbach.

ISSN 2364-08718
ISBN 978-3-631-80395-0 (Print)
E-ISBN 978-3-631-81513-7 (E-PDF)
E-ISBN 978-3-631-81514-4 (EPUB)
E-ISBN 978-3-631-81515-1 (MOBI)
DOI 10.3726/b16658

© Peter Lang GmbH
Internationaler Verlag der Wissenschaften
Berlin 2020
All rights reserved.

Peter Lang – Berlin · Bern · Bruxelles · New York · Oxford · Warszawa · Wien

All parts of this publication are protected by copyright. Any utilisation outside the strict limits of the copyright law, without the permission of the publisher, is forbidden and liable to prosecution. This applies in particular to reproductions, translations, microfilming, and storage and processing in electronic retrieval systems.

www.peterlang.com

To

Sandro & Finn
Thank you for the good moments in time as one family and for the bunches of love.

My Parents
Thank you for hanging in and supporting me warmheartedly in daily life, across this long-term project, from the first to the very last day.

Foreword

Military organizations see themselves as value-driven, and therefore it is not surprising that most of the armed forces have a definitive list of values and standards. This is a means of articulating its common ethos and moral basis, and ensuring correct behavior of serving personnel. Moreover, in an era where civilian society's values seem to be moving ever further away from those of armed forces such a basis becomes more relevant in educating soldiers on the moral character necessary to address the ethical situations they will encounter in military training and even more during military missions.

In general, we assume that values reflect what is accepted by cultures, be it the military organization or the wider population. However, it has not been proven that existing lists of values can be equated with it; they have mostly grown historically and are therefore based not least on traditional ideals, norms, standards, and also top-down guidelines. Robinson (2006) even goes so far as to say that lists of values and virtues produced by most armed forces in the Western world remain inwardly orientated and somewhat old fashioned (p. 266).

The Swiss Armed Forces have deliberately refrained from using an universal canon of values. But the official service regulations refer to certain values in the sections on leadership and education; the Military Ethics Report (2010) also stresses the importance of fundamental values; and finally, values are also part of the content of various leadership courses. On the one hand, this provides at least a certain framework, and on the other hand, the room for maneuver that is desirable in practice. But now one can argue that in a rapidly changing world and in view of our multioption society, specifically for a conscript army, it seems appropriate to create a more solid and binding basis in terms of content by systematically examining the values perceived.

This is precisely the starting point of this research project. Its aim was using scientific methodology to depict and interpret the current understanding of values and virtues; to intensify the dialogue on the culture of values and virtues within the military organization; and to create the conditions for evaluating the impact of military education. With this in mind, qualitative and quantitative data were collected at different hierarchical levels. The resulting comprehensive and representative data set opened up the possibility for comparisons between the different hierarchical levels and provided detailed information on the differences between professional members of the armed forces and members of the militia. Furthermore, a structural factor analysis of the same data set led to five superordinate value factors and four virtue factors, which somehow reflect the values-landscape of the Swiss Armed Forces. For example, the five value domains reflect a culture of respectful, human-oriented conduct and a simultaneous military focus on mission and related duties.

The benefits of the present study are manifold: In contrast to the traditionally established value systems of other military organizations, mostly imposed from above or by history, a value culture is described here from the perspective of today's members of the Swiss Armed Forces. In addition to this overall picture, it also allows comparisons between relevant subgroups and thus stimulates discussion about the different weighting of the values and virtues. It can also be highlighted that this research project should provide the impetus for self-reflection. Thus, while an officer is, first and foremost, a soldier, the value-oriented behavior of an officer goes above and beyond what is required of soldiers. So, the results presented here can be seen as a binding basis on which the behavior of leaders can be assessed and on which they should reflect themselves – their decisions and actions – regularly. And finally, this study can serve as a starting point for further scientific work such as the systematic recording of the effects of military value education or for comparisons of the understanding of values and virtues in an international context.

Nadine Eggimann Zanetti deserves credit for having had the courage to investigate this complex subject systematically and using scientific methods, thus making a substantial contribution to the research in question and providing a solid basis for practice in a subject of great importance not only for military organizations.

Hubert Annen

References

Robinson, P. (2006). *Military Honour and Conduct of War: From Ancient Greece to Iraq*. New York: Routledge.

Swiss Armed Forces. (2010). *Militärethik in der Schweizer Armee [Military ethics in the Swiss Armed Forces]*. Bern, Switzerland: Government Printing Office

Acknowledgments

I would like to express my thanks to all of you who have supported me throughout my doctoral dissertation.

First and foremost, I am extending my sincere thanks to Prof. Dr. Willibald Ruch for the opportunity to conduct this research and for all his guiding advice. His intensive support in exploring details, his academic expertise, and his wide-reaching supervision have enabled this project to deliver the essential findings, nuances, and new viewpoints in understanding values and virtues within the military context and beyond.

Furthermore, I would like to thank Prof. Dr. Bruno Staffelbach for agreeing to review my thesis and to let me benefit from his academic and military experience. I also thank him for his structured method of coaching me through the interdisciplinary stages finalizing the dissertation.

A very special thanks is addressed to Dr. Hubert Annen, my supervisor at the department of Military Psychology and Military Pedagogy Studies at the Military Academy at ETH Zurich (MILAK). His coaching and scientific directives helped me mature into what is required to cope with the diversity of challenges in creating value in academic and professional life. I gratefully appreciate all his commitment towards my project, his inspiring and insightful thoughts, and the ongoing flexibility to foster this project. Last but not least, I deeply value his trust and the friendship he shares with me all the way along.

Special thanks and sympathy go to Dr. Peter Stöckli and lic. phil. Can Nakkas, my professional colleagues and friends. We not only shared an office together for more than five years, but also went through ups and downs together, motivating and supporting each other, and not losing the proper doses of humor at times.

All the same thanks are shared with the whole team at the department of Military Psychology and Military Pedagogy Studies, with MSc Philippe Goldammer, MSc Madlaina Niederhauser, and MSc Regula Züger for their support in every way. In particular, I thank Philippe for his initiative in assisting me with the data collection for Study III.

Many thanks are due to the commanders of the Swiss Armed Forces who supported me with collecting data, conducting the military expert interview, and sharing their opinions on the subject of military values and virtues. The personal interaction with the participants was of great value to me. A very special thanks is conveyed to the former directors of the Military Academy (MILAK), Brigadier General (retired) Daniel Lätsch and Brigadier General Daniel Moccand, and to the current commander at the MILAK, Brigadier General Peter Stocker, for all their generous support and approval to conduct this research project.

I also would like to thank Colonel on the General Staff Reini Eugster for encouraging me to become a specialist officer in communication and press relations, thereby enabling me to experience military life in practice with all of its true meaning of comradeship, esprit de corps, loyalty, and discipline representing real military values and virtues.

An additional appreciation for sharing essential scientific, philosophical, and personal dialogue, especially during the initial phase of my dissertation, goes to Dr. theol. and Colonel on the General Staff Dieter Baumann. I would also like to thank Dr. Florian Demont and MSc Sabrina Pfister for sharing interdisciplinary initiatives within the domain of value research at the MILAK from the perspective of psychology, philosophy, and sociology.

Furthermore, I am expressing thanks to the recruits, soldiers, officer candidates, professional NCOs, and professional officers of the Swiss Armed Forces, who participated in my online studies on values and virtues.

I want to add my warm thanks to all my former and current colleagues for exchanging their thoughts with me as part of the doctorate colloquium, the PhD seminar, and the writing circle, both at University of Zurich, at the MILAK of ETH Zurich, and at University of Lucerne. In particular, I thank Dr. Angelika Güsewell and Dr. Sarah Auerbach for this intense and lasting PhD type of friendship.

I am very fortunate to also extend my thanks to Mrs. Gretchen Bain Matthews for the time-consuming and very valuable effort to proofread my dissertation.

A special thanks is conveyed to Dr. Priska Hubmann, who provided essential assistance in establishing the proper configuration for a professional MS word layout and literature references. Another warm thanks goes to MSc Regula Lätsch and to MSc Nicole Jehle for their valuable assistance as part of the data collection and data analysis.

And I do not want to miss extending my thanks to my family and friends in real life who created those special moments in time when I needed a good dose of balance, relaxation, and chats, adding additional strengths throughout the dissertation process.

Nadine Eggimann Zanetti

The Value of Values and Virtues
"Make us to choose the harder right
instead of the easier wrong
and never to be content with a half truth
when the whole can be won."

                        From the Cadet Prayer of the U.S. Military Academy

Contents

Management Summary 19

Zusammenfassung 23

Theoretical Background 27
- 1 Why study values and virtues in the Swiss Armed Forces? 27
  - 1.1 Characteristics of the Swiss military system 31
  - 1.2 In search of military core values and virtues in the Swiss Armed Forces 33
- 2 Positive psychology and how it applies to military psychology 38
  - 2.1 The scientific framework of positive psychology 38
  - 2.2 Good character: The revival of research on positive traits 40
    - 2.2.1 The VIA classification of strengths 42
    - 2.2.2 The VIA-IS 45
  - 2.3 Towards a positive military psychology 45
- 3 Research on values and virtues 49
  - 3.1 Values and virtues in philosophy 49
  - 3.2 Universal values: Definitions and theories 54
    - 3.2.1 Two main strains of value research 56
    - 3.2.2 Measurement instruments 66
    - 3.2.3 Values and personality 69
    - 3.2.4 Values and motivation 71
  - 3.3 Universal virtues: Definitions and theories 73
    - 3.3.1 Principle studies on identifying and structuring virtues .. 74
    - 3.3.2 The factor structure of character strengths and virtues ... 76
  - 3.4 The psycholexical approach towards the structure of values and virtues 78
    - 3.4.1 Psycholexical studies on universal values 79

		3.4.2 Psycholexical studies on universal virtues	81
	3.5	Classifications of values and virtues: Summarizing overview	82
	3.6	Conceptual difference between values and virtues	88
4	Research on military values and virtues		90
	4.1	Significant findings on the structure of military values and virtues	91
		4.1.1 Findings on military values	92
		4.1.2 Findings on military virtues	94
	4.2	International practical perspective	97
	4.3	Values and virtues in the Swiss Armed Forces	103
5	Research questions and aim of this present thesis		104
6	Procedure		108
	6.1	Pre-study: Development of the MVVC	109
		6.1.1 Psycholexical-oriented analysis	109
		6.1.2 Categorization in values and virtues	111
		6.1.3 Expert interview with high executive military leaders	112
	6.2	The MVVC applied in a hierarchical top-down data collection	114

Study I: Assessing the Structure of Military Values ... 119

7	The structure of military values and the relation to universal values and personality		119
	7.1	Introduction	119
		7.1.1 The psycholexical approach towards the structure of values	119
		7.1.2 Relations of values to personality	122
		7.1.3 Research on values in military psychology	123
		7.1.4 Aims of the study	125
	7.2	Method	125
		7.2.1 Participants and procedure	125
		7.2.2 Measures	126

	7.3	Results	128
		7.3.1 Primary analyses	128
		7.3.2 Factor structure of military values	128
		7.3.3 Five military value factors	131
		7.3.4 Relations of the five factors of military values to universal values and personality traits	133
	7.4	Discussion	136

Study II: Assessing the Structure of Military Virtues 141

8 The structure of military virtues and the relation to the five factors of the VIA-IS 141

- 8.1 Introduction 141
  - 8.1.1 Five factors of character strengths measured by the VIA-IS 142
  - 8.1.2 Research on character strengths and virtues among military samples 144
  - 8.1.3 The psycholexical approach towards the structure of virtues 145
  - 8.1.4 Aims of the study 146
- 8.2 Method 147
  - 8.2.1 Participants and procedure 147
  - 8.2.2 Measures 148
    - List of 42 military virtues as part of the MVVC 148
    - VIA-IS (Peterson, Park, & Seligman, 2005) 148
- 8.3 Results 148
  - 8.3.1 Primary analyses 148
  - 8.3.2 Factor structure of military virtues 149
  - 8.3.3 Four military virtue factors 152
  - 8.3.4 Relations of the four factors of military virtues to the five factors of character strengths 155
- 8.4 Discussion 156

Study III: Investigating the Criterion Validity of the Five Military Value Factors and the Four Military Virtue Factors 161

9   Can the military value factors and military virtue factors determine organizational citizenship behavior and motivation to lead? 161
    9.1   Introduction 161
        9.1.1   Factor structure of the MVVC 162
        9.1.2   OCB and MTL as validation criteria 164
        9.1.3   Aims of the study 168
    9.2   Method 168
        9.2.1   Participants and procedure 168
        9.2.2   Measures 169
                Independent variables 169
                Criterion variables 170
    9.3   Results 171
        9.3.1   Primary analyses 171
        9.3.2   Universal values, military values, and military virtues as determinants of OCB and MTL 173
        9.3.3   Incremental validity of military values and military virtues for determining OCB and MTL 175
    9.4   Discussion 176

General Discussion 181
  10   Main results and conclusions 181
  11   Strengths and limitations 185
  12   Implications for research and practice 188
  13   Open questions and further research 192
  14   Final comments 196

List of Figures 197

List of Tables 199

References 201

Management Summary

Expressions describing values and virtues (e.g., honesty, security, loyalty, integrity) have always been a top priority within the military context of leadership, training, ethical commitment, and psychological research. Values and virtues are qualified as morally good, positive characteristics of personality (De Raad & Van Oudenhoven, 2011). Likewise, concepts of positive psychology such as character strengths, values, and virtues are regarded as highly influential on work satisfaction, individual performance, motivation, adaptability, and effective leadership in the military context (Matthews, Eid, Kelly, Bailey, & Peterson, 2006b). Additionally, the military environment is a natural home for concepts of positive psychology and their overall importance is widely documented (Matthews, 2009). This bolsters the inclination to view the military, specially the Swiss Armed Forces, as a value-oriented organization. However, there has not yet been an empirical psycholexical consideration of military values and virtues and a corresponding analysis of the factorial structure within a military organization. Combining the method of a psycholexical approach with a factorial analysis as part of the structural assessment of values and virtues enables to identify cultural-sensitive characteristics of the organization.

The current thesis investigated military values and virtues within the Swiss Armed Forces, with the focus on a psycholexical-based identification of military value and virtue descriptors and the assessment of their factorial structure. Data were collected from professional and militia military persons across the entire hierarchical levels of the Swiss Armed Forces. The validity of the resulting military value and virtue factors was further explored by relating them to (a) measures of universal values by the Austrian Value Questionnaire (AVQ; Renner, Salem, & Alexandrowicz, 2004), (b) the Big Five personality traits (Big Five Inventory; John, Donahue, & Kentle, 1991), and (c) the five second-order factors of character strengths by the Values in Action Inventory of Strengths (VIA-IS; Peterson, Park, & Seligman, 2005). Furthermore, the positive outcome of military values and virtue factors was verified with regards to organizational citizenship behavior (OCB; Organ, 1997) and motivation to lead (MTL; Chan & Drasgow, 2001) among Swiss recruits.

As part of the Pre-study, the Military Values and Virtues Catalog (MVVC) was developed. The scope included the identification of 25 military value- and 42 virtue-descriptive expressions, evolving from a psycholexical-oriented analysis of military guidelines and in line with the prioritizing ratings as assessed by the top executives of the Swiss Armed Forces.

In Study I, the corresponding catalog was given to a sample of 550 career officers and career non-commissioned officers (NCOs) to capture the rating with regards to each specific military value and virtue as it applies to everyday military decisions and actions. Principal component analysis in combination with Goldberg's top-down approach delivered five military value categories (in terms of factors) reflecting the Swiss military culture, characterized as Freedom (I), Social Cohesion (II), Good Soldiership (III), Mutual Respect (IV), and Military Conformity (V). Relating the correlations with the Big Five personality traits (i.e., Neuroticism, Extraversion, Openness, Agreeableness, and Conscientiousness), the strongest ties were found between Good Soldiership (III) and Conscientiousness, as well as between Mutual Respect (IV) and Agreeableness. In regards to the five universal values (i.e., Intellectualism, Balance, Religiosity, Materialism, and Conservatism), Good Soldiership (III) showed the highest correlation to Conservatism, while the correlation coefficients were generally not higher than between military values and personality traits.

In Study II, the MVVC was administered to 270 militia officer candidates from different service branches, with the goal of assessing the structure of the 42 military virtue-describing terms. Applying the same methodological approach as in Study I, four military virtue factors were found: Fortitude (I), Suitable Behavior (II), Reflection (III), and Empathy (IV). The four-factorial structure of the military virtues showed a characteristic correlation pattern with the five second-order factors of character strengths (interpersonal strengths, emotional strengths, intellectual strengths, strengths of restraint, and theological strengths), highlighting that military virtues build upon the universal-related strength factors. Specifically, emotional strengths were strongly related to the four military virtue factors, among which the highest correlations related to Fortitude (I) and Suitable Behavior (II). Additionally, Fortitude (I) correlated negatively with strengths of restraint.

Study III included data obtained from Swiss military recruits ($N = 391$) undergoing basic military training. In detail, the objective was to analyze the effects of (a) universal values and (b) military values and military virtues on OCB and on the motivation to pursue a militia cadre career (MTL). In multivariate analyses, Intellectualism and Harmony, among universal values, as well as Good Soldiership (military value factor III) and Fortitude (military virtue factor I) proved to significantly determine OCB and MTL. Military values and virtues showed a larger criterion validity in reference to OCB than to MTL. Furthermore, an incremental contribution of military values and military virtues beyond the universal values with regards to the prediction of OCB and MTL was found.

Finally, 19 international military organizations were consulted, to inquire about the practical use of a classification of military values and virtues, and about the existence of similar research activities. The result was that 12 organizations confirmed that they apply the principles of core values and virtues, confirming the positive impact in fostering successful leadership.

The overall outcome of the thesis confirmed that the identification of the military-specific factors of values and virtues required a psycholexical and factor analytic approach, which allow to interpret the cultural-specific aspects of the results. The findings are providing implications for military education, training, and leadership and fundamental input for extended future research.

Zusammenfassung

Begriffe, die Werte und Tugenden beschreiben (z. B. Ehrlichkeit, Sicherheit, Loyalität, Integrität), haben im militärischen Bereich der Führung, der Ausbildung, der ethischen Grundsätze und der psychologischen Forschung traditionsgemäss eine hohe Priorität. Werte und Tugenden definieren sich als moralisch gute, positive Persönlichkeitsmerkmale (De Raad & Van Oudenhoven, 2011). Werte und Tugenden sind bedeutsame Konzepte der positiven Psychologie. Ihr Einfluss auf die Arbeitszufriedenheit, individuelle Leistung, Motivation, Anpassungsfähigkeit und erfolgreiche Führung hat sich ebenso bedeutsam im militärischen Kontext bestätigt (Matthews, Eid, Kelly, Bailey, & Peterson, 2006b). Entsprechend ist das militärische Umfeld vielseitig dokumentiert als ein natürliches Zuhause für Konzepte der positiven Psychologie (Matthews, 2009). Mit dieser Sichtweise wird die Schweizer Armee als eine werteorientierte Organisation betrachtet. Allerdings gab es bis anhin keine empirische psycholexikalische Bearbeitung der aktuellen militärischen Werte, Tugenden und deren strukturellen Zusammenhänge. Eine solche Kombination der psycholexikalischen Methode mit der Analyse der faktoranalytischen Struktur von Werten und Tugenden erlaubt eine spezifische Erfassung der Organisationskultur.

Dieses Forschungsprojekt untersuchte Werte und Tugenden in der Schweizer Armee. Der Schwerpunkt umfasste die psycholexikalische Identifizierung von militärischen Werte- und Tugend-beschreibenden Begriffen, sowie die Bestimmung ihrer faktoriellen Struktur. Die Daten wurden erfasst auf der Grundlage persönlicher Interviews und Online-Befragungen mit dem militärischen Berufs- sowie Milizpersonal über die gesamten Hierarchiestufen der Schweizer Armee. Die Validität der resultierenden militärischen Werte- und Tugendfaktoren wurde anhand von Korrelationen mit verwandten Konzepten verifiziert: (a) mit den universellen Werten (Österreichischer Wertefragebogen, ÖWF; Renner, Salem, & Alexandrowicz, 2004), (b) mit den Big Five Persönlichkeitsmerkmalen (Big Five Inventory, BFI; John, Donahue, & Kentle, 1991), und (c) mit den fünf Faktoren zweiter Ordnung der Charakterstärken basierend auf dem Values in Action Inventory of Strengths (VIA-IS; Peterson, Park, & Seligman, 2005). Ergänzend wurden die Zusammenhänge der militärischen Werte- und Tugendfaktoren mit organizational citizenship behavior (OCB; Organ, 1997) und Führungsmotivation (MTL; Chan & Drasgow, 2001) von Rekruten analysiert.

In der Vorstudie wurde der Military Values and Virtues Catalog (MVVC) erstellt. Dieser beinhaltete die Identifizierung von 25 militärischen Werte- und 42 Tugend-beschreibenden Ausdrücken, die sich auf der Grundlage einer psycholexikalisch-orientierten Analyse der militärischen Grundlagendokumente definierten und durch die obersten Führungskräfte der Schweizer Armee als prioritär eingestuft wurden.

In Studie I wurde der MVVC einer Stichprobe von 550 Berufsoffizieren und Berufsunteroffizieren vorgelegt, um die subjektive Präferenz für jeden militärischen Werte- und Tugendbegriff aus der Sicht von Entscheidungen und Handlungen im militärischen Alltag zu erfassen. Die Hauptkomponentenanalyse in Kombination mit Goldberg's Methode der hierarchischen Faktoranalyse definierte fünf militärische Wertefaktoren: Freiheit (I), Zusammenhalt (II), Good Soldiership (III), gegenseitiger Respekt (IV) und Hierarchie (V). Bezüglich den Korrelationen mit den Big Five Persönlichkeitsmerkmalen (Neurotizismus, Extraversion, Offenheit für Neues, Verträglichkeit, Gewissenhaftigkeit) wurde die stärkste Beziehung zwischen Good Soldiership (III) und dem Persönlichkeitsmerkmal Gewissenhaftigkeit, als auch zwischen gegenseitiger Respekt (IV) und dem Persönlichkeitsmerkmal Verträglichkeit ermittelt. Mit Bezug auf die fünf universellen Wertetypen (Intellektualismus, Balance, Religiosität, Materialismus, Konservatismus) zeigte Good Soldiership (III) die höchste Korrelation mit Konservatismus, während die Korrelationskoeffizienten generell nicht höher waren als zwischen militärischen Werten und den Persönlichkeitsfaktoren.

In Studie II wurde der MVVC 270 militärischen Offiziersaspiranten aus verschiedenen Truppengattungen vorgelegt, mit der Zielsetzung, die Struktur der 42 militärischen Tugend-Begriffen zu ermitteln. Unter Anwendung des gleichen methodischen Ansatzes wie in Studie I wurden vier militärische Tugendfaktoren ermittelt: Persönliche Stärke (I), korrektes Verhalten (II), Reflexion (III) und Empathie (IV). Die Vier-Faktoren-Struktur der militärischen Tugenden verwies auf ein charakteristisches Korrelationsmuster mit den fünf Faktoren zweiter Ordnung der Charakterstärken (interpersonelle Stärken, emotionale Stärken, intellektuelle Stärken, Stärken der Mässigung, theologische Stärken). Dies deutete darauf hin, dass militärische Tugenden auf den universell definierten Faktoren von Charakterstärken aufbauen. Insbesondere zeigten die emotionalen Stärken des VIA-IS die stärkste Verbindung mit den vier militärischen Tugendfaktoren, zu denen sich die höchsten Korrelationen in Bezug auf Persönliche Stärke (I) und korrektes Verhalten (II) identifizierte. Zusätzlich korrelierte Persönliche Stärke (I) negativ mit den Stärken der Mässigung.

Studie III umfasste eine Stichprobe von Schweizer Rekruten ($N = 391$), die sich in der militärischen Grundausbildung befanden. Von Interesse war es, die Auswirkungen von (a) universellen Werten und (b) den fünf militärischen Wertefaktoren und den vier militärischen Tugendfaktoren auf OCB und auf die Führungsmotivation (MTL) zu analysieren. Als Ergebnis der multivariaten Analysen erwiesen sich Intellektualismus und Harmonie unter den universellen Werten sowie Good Soldiership (militärischer Wertefaktor III) und Persönliche Stärke (militärischer Tugendfaktor I) als signifikante Vorhersage von OCB und MTL. Ausserdem zeigte sich, dass militärische Werte und Tugenden einen inkrementellen Beitrag zur Vorhersage von OCB und MTL erlaubten, ergänzend zu den universellen Werten.

Innerhalb einer ergänzenden internationalen Befragung wurden 19 militärische Organisationen konsultiert, um relevante Hinweise über die praktische Anwendung einer Werte- und Tugendklassifikation sowie über die Existenz ähnlicher Forschungstätigkeiten zu erfahren. 12 der Organisationen bestätigten eine aktive Anwendung von militärischen Kernwerten und -tugenden. Die Resonanz der Umfrage verwies auf deren thematische Bedeutung zu Gunsten der erfolgreichen militärischen Führung.

Das Ziel der Datenanalysen innerhalb dieser Dissertation war die Definition einer militärspezifische Klassifikation von Werten und Tugenden in Übereinstimmung mit der Kultur der Schweizer Armee. Als grundlegend dazu erwies sich die erstmalige Anwendung eines psycholexikalischen und faktoranalytischen Ansatzes im militärischen Umfeld zur Identifizierung der zentralen militärischen Werte und Tugenden sowie zur Bestimmung deren faktoriellen Struktur. Damit war ermöglicht, aufgrund eines systematischen Vorgehens die Kernwerte und -tugenden der Schweizer Armee zu definieren. Die Ergebnisse dienen als zusätzliche Grundlage für eine optimale militärische Erziehung, Ausbildung und Führung, und verweisen auf erweiterte zukünftige Forschung in Ergänzung zu dieser Dissertation.

Theoretical Background

1 Why study values and virtues in the Swiss Armed Forces?

> *"Instead of educating our soldiers, we train them. Instead of developing the male characteristics, as typically given to a good soldier, we let theory fill their heads."*
>
> — *General Ulrich Wille*[1] —

General Ulrich Wille, Commander-in-Chief of the Swiss Armed Forces in the First World War and head of the Swiss Military Academy at ETH Zurich from 1909 to 1913, considered values and virtues to be part of the "male characteristics," and as such subject to military education. He was the first military leader to introduce the subject of values and virtues into the practical operation and theoretical understanding of the Swiss Armed Forces (Annen, Steiger, & Zwygart, 2004). The wording of his statement may need to be adjusted to today's military operational reality, but its fundamental message remains valid. It is a historic point that values and virtues are understood as a binding part of military leadership, training, and education in the Swiss Armed Forces. Since the time of General Wille's command, all subsequent service regulations of the Swiss Armed Forces have included an explicit reference and commitment to military values and virtues.

This thesis devoted its research to the wide subject of values and virtues within the context of military psychology and the Swiss Armed Forces. The general concepts of values and virtues, and more explicitly, their content and implementation, are of great importance in military daily life. They are decisive in enabling a leader to execute his task thoughtfully, rather than harshly. The success of a training lesson in fostering the independent thinking of soldiers, rather than merely executing tasks, also depends on the values of the respective instructors. In both cases, values and virtues are implicitly noticeable.

According to Baumann (2007), values and virtues set standards and guidelines as to how to behave in a military environment. Annen et al. (2004) stated that values and virtues provide orientation and consistency to the individual and the entire military community. Furthermore, they are an important prerequisite to aim at the ultimate target regarding both education and conduct, as well

---

1 Wille, 1918, cited from Allgemeine Schweizerische Militärzeitschrift (1954, p. 106, "Erziehung und Wesen des Offiziers" ["Education and nature of an officer"]).

as to safeguard the ethical behavioral mandates under more severe conditions. Values and virtues as psychological concepts refer to stable characteristics of individuals, which are understood as positive, morally good, and desirable (De Raad & Van Oudenhoven, 2008). Accordingly, a value identifies what people find important and guides them in regards to choices and decisions. A virtue is generally understood to be a morally good trait, enabling a person to live in accordance with his or her personal values (De Raad & Van Oudenhoven, 2011). For instance, human dignity is interpreted as a value. Courage or bravery are virtues, which make it possible to live in accordance with human dignity, for instance, via a public engagement to the benefit of social minorities. This thesis strictly treated values and virtues as separate concepts, in line with this theoretical understanding.

Furthermore, the fundamental assumption within this thesis was that military organizations differ from civilian institutions in reference to their specific culture, its values, virtues, and traditions. Meyer (2015) described the specific difference as follows: "The military is, assuredly, a culture. It has its own history, laws, values, traditions, language, and customs" (p. 416). In line with Soeters, Winslow, and Weibull (2006), culture can be defined as common views on values and priorities in life. Consequently, culture is strongly linked to the concept of values. It represents a common vision of collectivities of people such as nations, regions, organizations, schools, churches, and families. Schein (1985) defined organizational culture as a set of assumptions, values, and beliefs that find shared acceptance by members of an organization. Accordingly, the military as an organization is assumed to have a specific organizational culture (Soeters et al., 2006).

The following characteristics contribute to calling the military a particular culture: *Soldiers*[2] wear uniforms, which makes them different from most other personnel in a highly visible way (Soeters et al., 2006). They receive their training in military schools and academies, where a sense of uniqueness is emphasized, assuming primacy of the group over the individual. Military persons normally work and sleep in separated barracks and bases and during this time military culture permeates nearly every aspect of their lives. As highlighted by Johansen, Laberg, and Martinussen (2013), veterans who served for only a few

---

2 The term *soldier* was used throughout the thesis in a generic sense and refers to any member of a military organization (including infantrist, airman, marine, etc.) of any military rank, addressing a *military person*. The meaning is independent of referring to a system of military conscripts (e.g., Switzerland) or military professionals (e.g., United States).

years reported strong identification with the military decades later. Even if the Swiss Armed Forces is not primarily involved in deployment and peace-keeping missions, active military duty can be dangerous and potentially life-threatening. As Druckman et al. (1997) pointed out, military organizations require a lot from their personnel. They are permanently on a 24-hour call and can be directed to remote locations at short notice.

Of course, different types of military organizations, such as the army, the air force, the navy, and the military police, have different cultures (Ross, Ravindranath, Clay, & Lypson, 2015). Each military branch has a number of values and virtues that military persons have to adhere to. Despite all this variation, there seems to be a general military-specific culture, as Soeters (1997) showed in his study. He compared military academies from 18 countries relating their view on various military values such as discipline and hierarchy. He found that military organizations from different nations were more similar to each other in reference to their value preferences than business organizations within the same nation of the respective military organization. Together with the evidence from other similar studies (e.g., Matthews, Eid, Kelly, Bailey, & Peterson, 2006b), it can be assumed that there is a distinct military culture that is different from civilian organizations.

Lang (1965) mentioned in his review three specific aspects which characterize the culture in military organizations: First, there is the communal character of military life, which describes the peculiarity that military and personal life often overlap, turning the job into a part of community life. As shown by Soeters (1997) in his study, military cultures proved to be far more institutional than corporate business cultures. In civilian business cultures leisure, personal life, and performance-based material gains are more preferred, while military cultures are more institutional, requiring a high commitment from their personnel, offering a fixed pay structure only. The second aspect observed by Lang (1965) referred to the importance of hierarchy, rules, and regulations in the organization. It may come as no surprise that military cultures are more hierarchical in comparison to the cultures of civilian enterprises (Soeters, 1997). Third, Lang (1965) mentioned the importance of discipline, which is described as the compliance with rules and as the acceptance of commands and authority. It can be further underlined through formal discipline (e.g., salutes, ceremonies, uniform) and functional discipline (e.g., acting in accordance with the rules and intent of the commander).

Above all, military culture can be characterized by a professional commitment that emphasizes discipline, hierarchy, courage, and self-sacrifice, setting the primacy of the group over the individual (cf. Collins, 1998; Hillen, 1999).

In other words, military institutions are legitimized in terms of specific values and virtues: a purpose, which transcends individual self-interest in favor of a presumed higher good. "Duty," "honor," "country," "courage," and "loyalty" are words that illustrate such *military values* and *military virtues*[3]. The guiding assumption within thesis was that military values and virtues can be used to describe and define the specific military culture (Kernic & Annen, 2016; Pathak, Rani, & Goswami, 2016).

It is not surprising that the military organization is a "value community," which identifies itself as a commitment to share common values (Moskos, 1973). This makes the military leaders accountable to explain, convey, and live the fundamental values of the organization. Such a commitment and leadership task is closely connected with the inherent concern of building an identity and a social cohesion (Kernic & Annen, 2016). In view of potential conflicts, a widely held postulate declares that the social competencies and psychological strength of the military members are a decisive factor in a mission's success or failure (Scales, 2009). In this respect, Matthews (2014) talked about "cognitive dominance," stating that considering psychological concepts such as character, values, and virtues will be of substantial importance for armed forces to be successful. For these reasons, it was recommended for a military organization to avoid the temptation to position leadership on principles, which are purely economic-purpose-rationally driven. Military organizations, specifically those rich in operational experience, care about character education, moral decision-making, and personal reflection on one's own values and virtues (Kernic & Annen, 2016).

Accordingly, values and virtues have always been a top priority within the military domain of leadership, training, ethical commitment, and psychological research. Military organizations have long recognized that morally good, positive characteristics of personality are highly influential on work satisfaction, individual performance, adaptation, and effective leadership (Matthews et al., 2006b). Specifically, the overall importance of character strengths, virtues,

---

3   The terms *military value* and *military virtue* do not have a consistent definition in literature and in military practice. This thesis was referring to the common understanding that military values and virtues are used as terms to characterize values and virtues of a military person. With this perspective in mind both wordings "military values and virtues" and "values and virtues of military persons" are equivalent. The present thesis used the terms *military values* and *military virtues* to describe military-specific values and virtues. However, this does not mean that military values and virtues cannot overlap with values of a civilian person; a military value such as trust can also be a guiding principle within the civilian environment.

and values within the military is widely documented (Matthews, 2009). Equally important, there is growing evidence that positive characteristics such as values, virtues, and character strengths predict success in challenging military situations. In spite of the emphasis in priority, there is still a dearth of empirical evidence, to demonstrate the importance of values and virtues within the military context (Matthews, 2012).

The Swiss Armed Forces is equally committed to foster military values and virtues such as discipline, comradeship, personal responsibility, as well as honesty (Swiss Armed Forces, 2004). Overall, the very specific environment of the Swiss Army represents a military institution with a long-lasting tradition, initiated by General Wille, to foster values and virtues, and expecting their members to respect them and live accordingly. The Swiss Armed Forces qualifies for representing a value-oriented organization (Proyer, Annen, Eggimann, Schneider, & Ruch, 2012). Like in other military organizations, the doctrine of the Swiss Armed Forces (Dienstreglement der Schweizer Armee, DR 04 [Swiss military Service Regulations 2004]) has emphasized the importance of personal values and virtues in successful leadership and military training (Annen et al., 2004).

1.1 Characteristics of the Swiss military system

The Swiss Armed Forces is an ideal environment to examine values and virtues given the distinct nature of the Swiss military system and the legal settings that govern the conscription of male citizens. The characteristics of the Swiss military system should be outlined for a general understanding of the samples of soldiers that are to be studied.

The main tasks of the Swiss Army are of defensive and protective nature. It also serves in case of natural catastrophes and other national hazards. The Swiss Armed Forces is a conscript army, in which all Swiss men aged between 19 and 31 years must fulfill their military service requirement. Active reserve officers serve even longer, until the age of 42 to 50 depending on their rank. The basic training lasts between 18 and 21 weeks followed by three weeks of training per year until the age of 26. Females may join the Forces voluntarily and are assigned to all groups including combat troops. Besides militia members, the Swiss Armed Forces employs a great variety of fulltime staff. However, a typical feature of the Swiss Armed Forces is that the greater portion of officers constitutes of members of the militia. Correspondingly, career officers, career NCOs as well as contracted military personnel account for only about 3% of the total Swiss Armed Forces (Annen, 2004).

Unlike the majority of the Swiss Armed Forces officers, who only serve as active reserve officers, career officers and career NCOs are fulltime professional officers. Prior to becoming a career officer or career NCO, the candidates completed at least one and a half years of training for militia officers or NCOs and were obliged to work temporarily as officers or NCO in the Swiss Armed Forces for one year, on average, to get practical experiences and a realistic insight into the actual job. The career personnel serve important functions such as educators and instructors, or preferably as coaches of the militia cadre. They are responsible for the molding of a leadership culture as well as the implementation of military guidelines.

The Swiss Army is a "training army," in which the professional officers and NCOs do not primarily have to be part of military operations and foreign assignments. However, the Swiss Armed Forces spends the majority of its time training for operational readiness and for educating soldiers, as part of their mandatory military service. Correspondingly, values and virtues are the basis of military education. Military education aims at influencing the values and the behavior of soldiers purposefully and sustainably (Annen et al., 2004). The Swiss Armed Forces with their militia system of a compulsory military service are regarded as a mirror of the society as a whole (Haltiner, 1996). Around 95% of Swiss military armed troops are conscripts, who serve as citizens in uniform (Szvircsev Tresch, 2011). They incorporate likewise the values and virtues of the Swiss society. The Swiss Armed Forces are thus obliged to present the objectives and the content of military education to both their military personnel as well as to the civilian society as a whole, in a transparent manner. That is why the values and virtues that characterize military education must be understood (Annen et al., 2004). Clarity about the conveyed values and virtues is needed, as they define the content of military education. For the military leader this means that it is required to understand the values, to live them as a role model, and to eventually embody them. In particular, the Swiss Report on Military Ethics published on September 1, 2010 (Swiss Armed Forces, 2010) stated that the training and educational culture of the Swiss Armed Forces must be explicitly defined according to values and virtues. Therefore, it was considered as indispensable to consciously deal with the soldiers' view and perception of values and virtues. The military education shall not convey a "counter world" to civil society (Eggimann & Annen, 2014). The Swiss Military Ethics Report is thus an up-to-date document, which refers to the importance of values and virtues in the Swiss Armed Forces and emphasizes the need for relevant scientific studies. However, currently there are no known military psychological studies, which have researched the values and virtues in the Swiss Armed Forces in a systematic and comprehensive manner.

1.2 In search of military core values and virtues in the Swiss Armed Forces

Matthews (2012, p. 214) stressed the need for a military organization to understand cultural significance as it applies to values and virtues, stating: "Cultural considerations are of paramount importance in twenty-first-century warfare." Furthermore, Britt, Adler, and Castro (2006) referred to the general requirement to address the question whether there are consistent values and virtues conveyed by the military organization. This perspective makes it a mandate to understand the core values and core virtues of an organization. *Core values* are principles that an organization views as being of central importance and reflect what the company values, setting the vision and goal of an organization (Duh, Belak, & Milfelner, 2010). Likewise, *core virtues* are the most important positive characteristics considered by an organization. Peterson and Seligman (2004) defined core virtues as "an abstract ideal, encompassing a number of other, more specific virtues that reliably converge to the recognizable higher-order category" (p. 35). Accordingly, core values and core virtues refer to categories of values and virtues which include expressions of thematically similar values and virtues grouped together[4]. Smolicz (1981) supposed that core values and virtues are forming fundamental components of a group's culture and have an identifying function with the group and its membership. Overall, core values and core virtues are the stated values and virtues, prioritized by a cultural group of persons, by a nation or by an organization[5]. They help define the culture of the organization, thus giving meaning to all its members (Pathak et al., 2016).

Britt et al. (2006) stated that it is essential to define a classification of values and virtues within the military organization. Such a *classification* is defined as a descriptive selection of one or many core values and virtues, indicating the most preferred ones as they apply to an organization (Albert, 1956). As Albert (1956) outlined, a corporate classification of core values and virtues within a military

---

[4] *Core values* and *core virtues* are derived from statistical value and virtue factors, which in themselves evolve from a structural analysis of an empirical data set representing subjective ratings of preference. Accordingly, value and virtue factors are categories of values and virtues, identified by a descriptive name of the respective groups of expressions that all have the similar meaning, e.g., integrity as a possible virtue factor implies a group of individual virtues such as sense of responsibility, reliability, discipline considerateness, and faithfulness.

[5] Although military values and virtue were captured on an individual basis within this thesis, the understanding of military core values and virtues evolved on an organizational level.

organization is conceived as a representation of the cultural organizational consensus, envisaged as a point of reference for the description and reflection of individual differences in values and virtues.

So far, there has not been a valid classification of core values and virtues reflecting the culture of the Swiss Armed Forces. Each military leader gets the freedom to prioritize his or her own personal values and virtues, independent of a binding corporate classification of core values and virtues. However, the Swiss Report on Military Ethics (Swiss Armed Forces, 2010) makes it an important priority to devote scientific effort to further assess the values and virtues of the Swiss Armed Forces. Such a commitment is in support in defining which values and virtues are to be fostered across the Swiss Armed Forces and conveyed to the Swiss soldiers as part of the military education. The benefit results from providing a common corporate understanding of values and virtues within the Swiss Armed Forces (Annen, 2017). Furthermore, a valid classification allows for a transparent communication within the framework of military education and provides the opportunity to prove the impact of military education and value-based leadership. Specifically, it needs detailed assessment to understand which core values and virtues, e.g., Mutual Respect or Fortitude, are being propagated in order to evaluate the efforts of military education. Clarity on value and virtues provides the condition for military leaders to apply a discipline of self-reflection and interactive dialogue (Eggimann & Annen, 2014).

In spite of a high regard for the meaning of values and virtues by the military organization, the empirical approach in research has been falling short (Schumm, Gade, & Bell, 2003). However, outside of the military domain, there has been an increasing interest in studies identifying and structuring taxonomies of universal[6] values and virtues. Specifically, Aavik and Allik (2002) gave preference to the psycholexical approach to develop a comprehensive and culture-sensitive list of universal values. De Raad and Van Oudenhoven (2011) initiated the same approach in classifying and structuring virtues followed by a series of corresponding research. The question of how many universal core values and core virtues[7] can be distinguished has been investigated through a psycholexical and factor analytic analysis in a variety of different cultures (e.g., Aavik & Allik, 2002; De Raad & Van Oudenhoven, 2008, 2011; Morales-Vives, De Raad, & Vigil-Colet, 2012, 2014; Renner, 2003b).

---

6   *Universal values* are considered to be generically applicable, applying to the broad civil society and not to a specific context such as the military.

7   *Core value* is a common expression in the military-specific literature (Pathak et al., 2016). Likewise, in the domain of universal virtues, *core virtue* is known from the work by Dahlsgaard, Peterson, and Seligman (2005) and Peterson and Seligman (2004).

With this theoretical perspective in mind, the thesis represented an initial comprehensive effort to apply the psycholexical approach in assessing the structure of values and virtues in a military organization. As mentioned previously, this research implied the assumption that the military culture differs from the civilian environment. Accordingly, it was of significant interest to capture the uniqueness of the military value and virtue culture of the Swiss Armed Forces. The psycholexical method was described as sensitive to culture-related differences (De Raad & Van Oudenhoven, 2008). Worth mentioning, that in the Swiss Armed Forces there is still no comprehensive description and classification which reflect the cultural-specific aspects of the military organization and the views of the different military subgroups (e.g., including military militia and military professionals). Moreover, there are numerous value- and virtue-descriptive expressions such as responsibility, loyalty, security, and freedom used in the official and inofficial Swiss military documentation (Baumann, 2007), which need further structuring for effective military leadership. It was essential to verify the relations between the expressions at scientific level, to group them according to their meaning and relevance, and to reduce to a manageable number. Finally, the identification and structuring of the essential military value- and virtue-describing terms led to the definition of the core values and virtues reflecting the military culture.

In summary, there is increasing recognition that values and virtues are positive characteristics of soldiers and important regarding military leadership, training, and education. At the same time, there is limited empirical conclusion as to which values and virtues characterize a military organization (Soeters, Poponete, & Page, 2006). The research within this thesis referred to the Swiss Armed Forces and extended the current general understanding of military values and virtues. The overall scope focused on identifying the relevant expressions of military values and virtues by means of a psycholexical approach and on assessing the structure concerning the military core values and virtues of the Swiss Armed Forces.

The thesis is divided into a Pre-study and three empirical parts, aiming at (a) the psycholexical-based identification of the Swiss military values and virtues and at establishing a corresponding catalog; (b) assessing the factorial structure of military values and virtues; (c) analyzing the relationship to measures

---

In the research domain of universal values, however, the expression "basic value" is more common (De Raad & Van Oudenhoven, 2008). This thesis was consistently using the term *core values* and *core virtues*.

of universal values, Big Five personality traits and factors of character strengths (Peterson & Seligman, 2004); and (d) exploring the criterion validity of the identified military value and virtue factors with regards to organizational citizenship behavior (OCB; defined by Organ, 1997, as the willingness to do more than what is normally demanded) and motivation to lead (MTL; defined by Chan & Drasgow, 2001, as the person's efforts to assume leadership training, roles, and responsibilities).

Figure 1 provides an overview of the Pre-study and the three empirical studies.

As overviewed in Fig. 1, the selection of the value and virtue descriptors in the Pre-study was conducted in three stages, i.e., (a) psycholexical search of existing military documentation, (b) consultation of military psychologists, and (c) interviews with high ranking military commanders of the Swiss Armed Forces. This approach delivered a valid and comprehensive list of 25 military value- and 42 virtue-descriptive terms, called the MVVC. The catalog was used to assess military values and virtues in the subsequent three empirical studies.

In Study I, a sample of Swiss career officers and career NCOs was tested to capture the factor structure of military values. The objective was to make the MVVC and its corresponding items subject to a factor analytic analysis, and to conclude on the core values in the Swiss Armed Forces. Given the fact that a military organization reflects a specific culture, it was of interest to evaluate how the military value factors correlate with universal values and Big Five personality traits. This added verification as to how the outcome of the military value factors compare with the universal factorial structure (De Raad & Van Oudenhoven, 2008, 2011).

Study II applied a similar approach as Study I, focusing on the structure of military virtues to be assessed in a large group of Swiss officer candidates. Additionally, the aim of Study II was to assess how the military virtues relate to the five factors of character strengths as measured by the VIA-IS (Peterson, Park, & Seligman, 2005).

As a result of the factor analytic analysis in Study I and II, the number of 25 military values and 42 virtues was reduced to five military value factors and four virtue factors[8]. To sum up, the prime aim of assessing the structure of military values (Study I) and of military virtues (Study II) was to facilitate a deeper understanding of the nature of the military culture.

---

8   Within this thesis, the terms *factor* and *dimension* are understood as equal in meaning, being defined as a group or category of similar values and virtues intercorrelating and derived by factor analysis. For consistency in terminology, the term *factor* was used throughout.

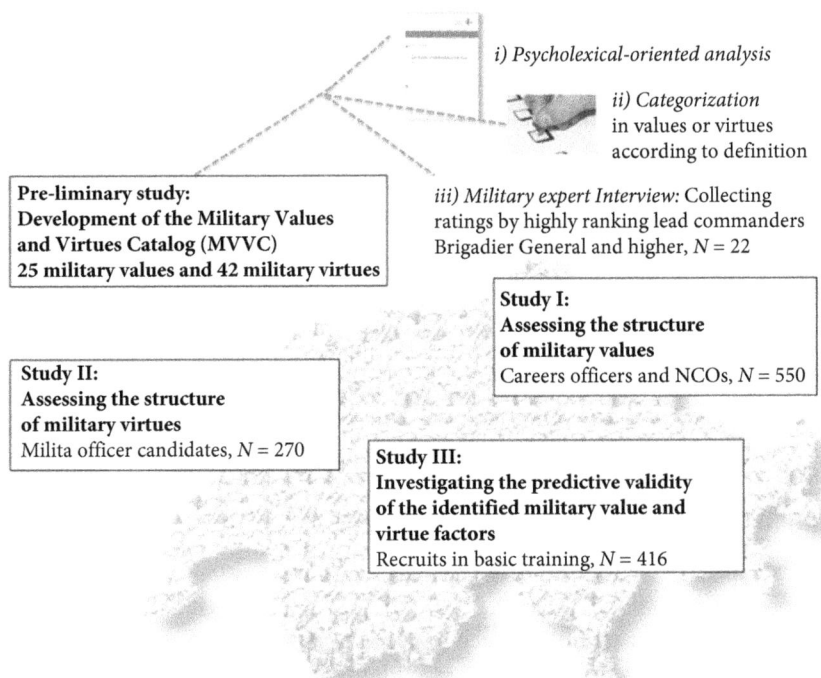

Fig. 1: Summary of the Pre-study and the three studies within the Swiss Armed Forces

Study III concerned the question whether the military value and military virtue factors exhibit criterion validity with regards to OCB and MTL. OCB and MTL were confirmed to be crucial concepts for a successful selection of Swiss military officers (Annen, Goldammer, & Szvircsev Tresch, 2015). Furthermore, previous research has shown that values and virtues relate to a person's motivation to accept a leadership role (Clemmons & Fields, 2011) and to display OCB (Halbesleben, Bolino, Bowler, & Turnley, 2010).

Overall, the thesis contributed to research regarding the impact of military values and virtues on training, leadership, and organizational structure within the Swiss Armed Forces. The special focus lay on the identification of the factorial structure of Swiss military values and virtues by means of a psycholexical and factor analytic approach.

This dissertation is composed of a theoretical background, three chapters of results which concerns the factorial structure of military values and virtues, and a general discussion. The first section includes an introduction to positive psychology, referring to Peterson and Seligman's (2004) concept of character strengths and the corresponding application of this theoretical framework to the traditional framework of military psychology. The subsequent section addresses the philosophical notion as well as the psychological approaches to universal values and virtues, the corresponding measurement instruments, and the research assessing the related factorial structure. Further studies describing values and virtues in the military context are highlighted in the following chapter including the emphasis in relevance within the Swiss Armed Forces. Next, the research questions as well as the specific aspects of the methodology and procedure are discussed. The subsequent three chapters describe the three studies that were conducted within the scope of this doctoral dissertation. The thesis ends with a general discussion highlighting key findings, outlining added value and limitations, and addressing implications for practice and extended research.

2 Positive psychology and how it applies to military psychology

First, an introduction is given to the science of positive psychology with specific reference to the notion of good character, to which positive characteristics such as values and virtues are linked. Second, the emerging field of positive military psychology is addressed and empirical evidence on the benefits of good character in the military setting is summarized.

2.1 The scientific framework of positive psychology

As mentioned above, values and virtues are assumed to be morally good and desirable characteristics of personality (De Raad & Van Oudenhoven, 2008), and therefore are an integral part of the overall positive psychological concept. Positive psychology is an umbrella term for theories and research about what makes life most worth living (Peterson & Park, 2003; Seligman, 2002). A preference for the scientific approach has emerged to systematically study positive characteristics, positive emotions, and positive institutions. During the first half of the 20th century, psychology pursued the following three distinct aims: "curing mental illness, making the lives of all people more productive and fulfilling, and identifying and nurturing high talent" (Seligman & Csikszentmihalyi, 2000, p. 6). After the Second World War, the field of psychological research had reduced its scope to only one of these missions, specifically curing mental illness. For the subsequent years, psychology was dominated by a disease model of human behavior (Seligman, 2000). The focus was now largely on pathology, on the negative effects of environmental stressors, and on the assessment and treatment of psychological disorders. Topics like the normal functioning of human beings, the application of personal strengths, and experience of positive emotions were not subjects of key interest (Gable & Haidt, 2005; Seligman, 2000). In other words, psychology as a discipline has done very little to support the majority of the population, who are healthy and psychopathology-free, to live the psychological "good life" (Seligman & Csikszentmihalyi, 2000, p. 5), meaning to become more productive and successful, and to develop a sense of positive engagement and meaning in life. This is what the science of positive psychology is destined to pursue.

Positive psychology was launched in 1998 by Martin Seligman during his term as president of the American Psychological Association (APA). One of his presidential initiatives was to bring forward the term "positive psychology," to promote systematic research on flourishing individuals and thriving communities in order to learn how to foster happiness, and life and work satisfaction (Seligman, 1998). One of the main theoretical precursors is humanistic psychology. This psychological movement in the 1960s and 1970s assumed human beings to have an innate need to strive for personal growth, fulfillment, and satisfaction in life as a basic human motive. Whereas the practitioners of humanistic psychology were skeptical about scientific method, the positive psychology movement stated that "both strength and weakness" (Peterson & Seligman, 2004; p. 4) could be empirically studied. However, positive psychology is not intended to replace traditional models and methods that psychologists employ in their practice or research. It is meant to balance the positive and negative aspects of life and to empirically study human flourishing, covering the full range of what makes human life (Gable & Haidt, 2005). Specifically, the research of positive psychology centers on three topics (Peterson, 2006; Seligman & Csikszentmihalyi, 2000):

(a) Positive subjective experiences (e.g., happiness, flow, pleasure)
(b) Positive individual characteristics (e.g., security or honesty as a value, courage or wisdom as a virtue, self-regulation or humor as a character strength)
(c) Positive institutions (e.g., families, workplaces, schools), which should enable the display of positive characteristics, like values, virtues, and character strengths, and which in turn foster positive experiences.

Positive psychology has grown rapidly in the last 18 years and now involves hundreds of researchers in the USA and all over the world. Much research has been conducted ever since to understand the factors and processes, which enable individuals and communities to lead the psychological good life (Seligman & Csikszentmihalyi, 2000).

2.2 Good character: The revival of research on positive traits

A further domain of research within positive psychology concerns the identification, measurement, and cultivation of good character as an expression of positive traits (Park & Peterson, 2009, 2010; Peterson & Seligman, 2003). Accordingly, a classification of virtues and character strengths was developed, intended to serve as a counterpart to the Diagnostic and Statistical Manual of Mental Disorders (DSM; American Psychiatric Association, 2000). Peterson and Seligman (2004) introduced the VIA classification of strengths as a framework for the investigation of character, virtues, and character strengths (details on the VIA classification will be outlined in section 2.2.1).

The notion of character traces back to ancient philosophy, referring to a mark impressed upon a coin. Specifically, it originates from the Greek word *charassein*, which means to scratch or engrave. Accordingly, the understanding of character has a long history. Following Aristotle's Nicomachean Ethics (Aristotle, trans. 2000) the concept of character implies traditionally a variety of personal attributes to live a morally good life. Similarly, Aristotle and other early Greek philosophers such as Socrates, Plato, Augustine, and Aquinas see character as the way to make someone a good person (Peterson & Seligman, 2004).

As compared to personality, character can be modified and developed with changing life circumstances and training activities (Peterson & Seligman, 2004). According to James (1899), character is the internal habit of thoughts, feelings, and action that everyone develops and that results in ultimate, authentic success. He saw the main task of a teacher as character building and understood character to be defined in the form of habits: "Your task is to build up a character in your pupils; and character, as I have so often said, consists in an organized set of habits of reaction" (James, 1899, p.108). Furthermore, he stated that "the (…) 'character' of an individual means really nothing but the habitual form of his associations. To break up bad associations or wrong ones, to build others in, to guide the associative tendencies into the most fruitful channels, is the educator's principal task" (p. 57–58). Consequently, character is assumed to be capable of adaptation as a result of repetitive habits.

There is a broad variety of definitions on character as part of a complex positive psychological concept. Lickona (1991) sees character as doing the right thing despite outside pressure to the contrary. Furthermore, Berkowitz (2002) defines character as an individual set of psychological characteristics that affect the person's ability to function morally. Pawelski (2003) summarizes a form of global definition of character stating that it is comprised of those characteristics that lead a person to do the right thing or not to do the right thing. Indeed, a common aspect of the various theoretical understanding of character is that it emphasizes volition and morality (Saucier & Srivastava, 2015).

The initial study of morality and character within personality psychology can be seen with Gordon Allport (e.g., Allport & Allport, 1921). Allport (1937, p. 51) assumed that when "personal effort is judged from the standpoint of some code" that is based on social standards it is called character. Accordingly, Allport also (1937, p. 52) stated that "character is personality evaluated." He sought to exclude ethical judgments from personality research. For Allport, the more evaluative the term is, the less reference to personality exists and the less value for the psychologist is gained. He considered such an ethical perspective on personality not necessary for psychology. Indeed, during the period of Allport's greatest influence, the use of the term "character" became uncommon in personality psychology. Overall there was a trend to have the terminology of character substituted by personality representing the inclusion of biophysical and psychological characteristics.

With the emergence of positive psychology at the beginning of the 21st century, the notion of character was taken up again as the inward determinant of a good life (Peterson, 2006). Specifically, Peterson and Seligman (2004) assumed that character is plural, not singular, and construed as a set of positive traits such as virtues and character strengths. Park and Peterson (2009, p. 1) refer to the importance of character with these significant words: "Good character is what we look for in leaders, what we look for in teachers and students, what we look for in colleagues at work, what parents look for in their children, and what friends look for in each other." They continue by saying that good character "is not the absence of deficits and problems but rather a well-developed family of positive traits." Peterson (2006) defines character as a family of individual differences, in principle distinct strengths that people possess to varying degrees, shown in thoughts, feelings, and actions. According to Boe (2017), a person can express his or her values through one's character as a correlate of positive traits such as virtues. This understanding points to the fact that character and values are linked to each other. The VIA classification by Peterson and Seligman (2004) describes good character on three conceptual levels, where virtues as moral character traits constitute the highest level. Accordingly, good character is used primarily in relation to virtues (see section 3.6 on the conceptual difference between values and virtues).

It is worth mentioning that this new understanding of character within the framework of positive psychology relies on the notion of personality psychology, which considers individual differences to be stable, but also shaped by the individuals' setting, and therefore subject to change. According to Pawelski (2003), the new approach will lead to important answers regarding the issues of how character might be assessed.

2.2.1 The VIA classification of strengths

In the context of establishing positive psychology, Peterson and Seligman (2004) argued that it was necessary to develop a classification of positive traits (virtues). With this primary objective in mind, they conceptualized a "hierarchical classification of positive characteristics" (p. 13) to categorize, define, and measure important character strengths. The project resulted in the *VIA classification of strengths*, which describes good character[9] via the following three conceptual levels:

(a) virtues,
(b) character strengths, and
(c) situational themes (from the highest to the lowest level).

Table 1 provides a list of the virtues and character strengths including their definitions.

As illustrated in Tab. 1, virtues I–VI represent the level of (a) six virtues. Within this context, Peterson and Seligman (2004) relied on the six core virtues described by Dahlsgaard, Peterson, and Seligman (2005) which have demonstrated a continuing relevance in literature and across different cultures. Peterson and Seligman (2004) argued that these core characteristics are cross-cultural, widely recognized, and ubiquitous, suggesting the possibility of being universal. It has even been suggested that they are rooted in biology through evolutionary processes. However, since the virtues are understood at a level that is rather abstract and general, the authors did not intend to measure the concept of virtues. They focused on assessing the level of the (b) 24 character strengths [(1) to (24)], which represent the components of good character as measurable individual differences[10]. To establish a list of character strengths, Peterson and Seligman (2004) collected entries for "psychological ingredients – processes

---

9   Within the VIA classification, a good character is constituted in degrees rather than in categories (McGrath, Rashid, Park, & Peterson, 2010; Peterson & Seligman, 2004). Peterson and Seligman (2004) supposed that a good character might be described by all virtues being present with "above-threshold values for an individual" (p.13), implying a similar meaning as when Park and Peterson (2009) stated "character is plural rather than singular" (p. 3).

10 Although the 24 character strengths are supposed to be universally valued, nobody is expected to demonstrate them all. It is rather assumed that someone has a good character if he or she displays one or two strengths within each of the virtue groups and that every individual has a distinct character strengths profile, that is a specific rank order of the strengths, from the most to the least central.

Tab. 1: *Classification of the six core virtues and 24 character strengths (Peterson & Seligman, 2004, pp. 29–30)*

Virtue I.	Wisdom and knowledge: intellectual strengths that entail the acquisition and use of knowledge
(1)	creativity [synonyms are originality, ingenuity]: thinking of novel and productive ways to do things
(2)	curiosity [interest, novelty-seeking, openness to experience]: taking an interest in all of ongoing experience
(3)	open-mindedness [judgment, critical thinking]: thinking things through and examining them from all sides
(4)	love of learning: mastering new skills, topics, and bodies of knowledge
(5)	perspective [wisdom]: being able to provide wise counsel to others
Virtue II.	Courage: emotional strengths that involve the exercise of will to accomplish goals in the face of opposition, external or internal
(6)	bravery [valor]: not shrinking from threat, challenge, difficulty, or pain
(7)	persistence [perseverance, industriousness]: finishing what one starts
(8)	integrity [authenticity, honesty]: speaking the truth and presenting oneself in a genuine way
(9)	vitality [zest, enthusiasm, vigor, energy]: approaching life with excitement and energy
Virtue III.	Humanity: interpersonal strengths that involve "tending and befriending" others
(10)	love: valuing close relations with others
(11)	kindness [generosity, nurturance, care, compassion, altruistic love, "niceness"]: doing favors and good deeds for others
(12)	social intelligence [emotional intelligence, personal intelligence]: being aware of the motives and feelings of self and others
Virtue IV.	Justice: civic strengths that underlie healthy community life
(13)	citizenship [social responsibility, loyalty, teamwork]: working well as member of a group or team
(14)	fairness: treating all people the same according to notions of fairness and justice
(15)	leadership: organizing group activities and seeing that they happen
Virtue V.	Temperance: strengths that protect against excess
(16)	forgiveness and mercy: forgiving those who have done wrong
(17)	modesty and humility: letting one's accomplishments speak for themselves
(18)	prudence: being careful about one's choices; not saying or doing things that might later be regretted
(19)	self-regulation [self-control]: regulating what one feels and does
Virtue VI.	Transcendence: strengths that forge connections to the larger universe and provide meaning
(20)	appreciation of beauty and excellence [awe, wonder, elevation]: noticing and appreciating beauty, excellence, and/or skilled performance in all domains of life

(continued on next page)

Tab. 1: Continued

(21)	gratitude: being aware of and thankful for the good things that happen
(22)	hope [optimism, future-mindedness, future orientation]: expecting the best and working to achieve it
(23)	humor [playfulness]: liking to laugh and joke; bringing smiles to other people
(24)	spirituality [religiousness, faith, purpose]: having coherent beliefs about the higher purpose

and mechanisms – that define the virtues" (Peterson & Seligman, 2004, p. 13) and then evaluated them using various methodological methods (e.g., review of literature on good character, brainstorming in core groups of scholars, and analysis of American Boy Scouts and of popular song lyrics). Moreover, a list of several defining criteria of character strengths was used to reduce the initial list of human strengths. For instance, a criteria for a character strength is "fulfilling" (i.e., contributing to individual fulfillment, satisfaction, and happiness broadly) and "measureable" (i.e., having been successfully measured by researchers as an individual) (cf. Park & Peterson, 2007). Additionally, it was hypothesized that strengths are "distinguishable routes to displaying one or another of the virtues" (Peterson & Seligman, 2004, p. 13).

The lowest level of the VIA classification is defined by the (c) situational themes. These are "specific habits that lead people to manifest given character strengths in given situations" (Peterson & Seligman, 2004, p. 14). For instance, zest might be shown in a different way at work, within the family, or in a group of peers. Zest at work may manifest in broad engagement and interest in the topics relevant at work, but zest within a peer group can be displayed differently, such as organizing special events and meetings to be together. However, there are fewer research studies on situational themes than on character strengths.

Ruch and Proyer (2015) empirically verified the structural model of the VIA classification by including expert judgments. Participants were instructed to rate each strength to the extent of how prototypically it corresponds with a virtue. Results within this study supported the structure suggested by Peterson and Seligman (2004). Furthermore, they also showed that the assignment of the strengths to virtues was confirmed, as theoretically proposed by Peterson and Seligman (2004), with humor as the only strength that better fits to the virtues of humanity or wisdom than to the virtue of transcendence.

From the viewpoint of this research, the importance of the VIA classification is twofold: First, it provides a well-established framework to classify and systematically assess universally valued positive characteristics. Virtues are seen as moral character traits (De Raad & Van Oudenhoven, 2011) and good character is a function of the six virtues and 24 character strengths. Accordingly, virtues are measured on the level of character strengths. Second, the hierarchical organization in different categories suggests which character strengths are similar and which are not. This provides the framework from which an individual profile of character strengths can be generated, and the components of good character can be assessed.

2.2.2 The VIA-IS

To measure the 24 character strengths, several instruments have been created. The established instrument for measuring character strengths is the VIA-IS (Peterson et al., 2005)[11]. It is a self-report questionnaire (10 items per strength) with 240 items using a 5-point Likert-scale (from 1 = *very much unlike me* through 5 = *very much like me*). The mean of the 10 items of each scale calculates the scale score. Validation was based on the data of over 150,000 adults: Peterson and Seligman (2004) reported substantial Cronbach alphas of all scales ($\alpha > .70$) and satisfactory test-retest correlations for all scales over a 4-month period ($> .70$). Some small relations to demographics were found. For example, women had higher scores in the strengths of humanity than men, younger adults scored higher in humor than older ones, and married participants rated themselves higher on forgiveness than divorced ones.

The original version of the VIA-IS is in the English language and was developed in several steps (Peterson & Seligman, 2004). Ruch et al. (2010) adapted the VIA-IS into German. As reported in this study, internal consistencies of the German version ranged from .71 (integrity) to .90 (spirituality), with a median of .77. Retest reliabilities were equivalent to the internal consistencies. Relationships of the German VIA-IS with demographics were modest but meaningful, and comparable to the ones found for the original VIA-IS. It is this German version of the VIA-IS by Ruch et al. (2010), which was applied in the Study II of this thesis.

2.3 Towards a positive military psychology

Matthews (2008) is acknowledged as the initiator for anchoring the connection of the military with positive psychology in a first publication. He summarized

---

11 Other instruments are the Values in Action Inventory of Strengths for Youth (VIA-Youth; Ruch, Weber, Park, & Peterson, 2014); the Values in Action Structured Interview (VIA-SI; Peterson, 2003); Values in Action Rising to Occasion Inventory (VIA-RTO; Peterson & Seligman, 2004); and the Brief Strengths Test (BST; Park & Peterson, 2007).

the military studies on positive psychology so far, and thus introduced the concepts of positive psychology into the military organization. This had come at a time when military psychologists were faced with the increasing consequences and human challenges of the US military's lengthy combat operations in Afghanistan and Iraq with a large number of soldiers[12] and veterans suffering from posttraumatic stress disorder (PTSD). It also had become difficult to select and prepare new soldiers for combat exposure and its psychological risks. Thus, the time had come for a paradigm shift in military psychology, adopting new ways of practice and research. According to Matthews (2008), positive military psychology is not considered to replace traditional models and methods that military psychologists apply in practice and research. Rather, positive psychological concepts and methods are proposed as a supplement to the military psychologist's toolbox. More precisely, it is argued that the military is a perfect "home" for concepts of positive psychology such as character strengths, values, and virtues (Matthews, 2008). A military environment is composed of relatively young, healthy, and pathology-free individuals (cf. Booth et al., 2007). Moreover, Matthews (2009) argued that the military is seen as a positive institution, qualifying for an organization that offers main services to society (i.e., as education and training of young men becoming soldiers, contributing to national security) becoming a key resource for collectivity. Overall, the military is an institution that works for the greater good of a society, with a strong emphasis on character development, values, morale, and welfare (Matthews, 2009).

Values, virtues, and character strengths are recognized as being critical for military leadership (Matthews et al., 2006b). There are a number of studies showing evidence that positive personality traits and good character predict success, effecting leadership, coping, and adaptation in challenging military contexts. In the following a set of studies is presented to illustrate the extending research devoted to positive psychology within military psychology in order to analyze the role of positive characteristics of soldiers.

The first study on applying positive psychology to the military was on "grit," a positive character trait defined as a measure of passionate pursuit of long-term goals. Duckworth, Peterson, Matthews, and Kelly (2007) looked at how grit might have a contribution in predicting retention in Cadet Basic Training (CBT) and for academic performance in the first year at the United States Military Academy

---

12 The term "soldier" is used throughout the thesis in a generic sense and refers to any member of a military organization (including infantrist, airman, marine, etc.), addressing a military person.

("West Point"). They compared the results with alternative predictors such as aptitude, leadership, and physical fitness. As they reported in their study, grit was the only statistically significant variable in predicting the successful retention in CBT. Also, no evidence was given that grit correlates with aptitude, leadership, or physical fitness measures. Additionally, grit was a significant predictor of academic grades during the cadet's first year at West Point. Matthews, Peterson, and Kelly (2006a) had all incoming members of West Point rate themselves on the 24 character strengths. At the end of CBT they compared the mean self-ratings in the VIA-IS and concluded that cadets who successfully completed CBT rated themselves significantly higher than those who left on nine strengths: bravery; vitality (zest); fairness; integrity; persistence (according to Matthews [2012], a trait highly correlated with grit); hope/optimism; leadership; self-regulation; and citizenship/teamwork. It is interesting to mention that these nine strengths are represented in the military doctrine and are therefore understood as relevant to soldier performance. It became evident that there is an overlap between the seven "Army Values" (loyalty, duty, respect, selfless service, honor, integrity, and personal courage; US Department of the Army, 2006) and the nine character strengths related to successful completion of West Point CBT. In conclusion, the results of this study pointed out that positive characteristics such as values, virtues, and character strengths are relevant to describe good character, successful leadership, and soldier performance in the military context.

Another important study, which exemplified how concepts of positive psychology are well suited to assess the nature of good character of soldiers and the overall military, was the study by Matthews et al. (2006b). They compared the VIA-IS-assessed character strengths of a sample of West Point cadets with two comparison groups of Royal Norwegian Naval Academy cadets and US civilians. The results showed that the two military samples consisting of young men and women attracted to military service manifested a different profile of the 24 strengths compared to the civilian counterparts. More precisely, the West Point cadets were more similar in their rank ordering of character strengths to Norwegian cadets than they were to their own fellow American citizens. Furthermore, equivalent character strengths seemed to be important for military success in both samples of West Points and Norwegian cadets. These results allowed for two interpretations. First, military culture is more influential in shaping character strengths of soldiers than the difference in national origins might suggest, and second, that the military environment attracts persons with similar profiles in character strengths.

Additionally, another set of field studies allowed relating positive characteristics to a variety of aspects of soldier adaptation and performance in training exercises. Particularly interesting with direct relation to virtues was the study by Eid, Matthews, and Johnsen (2004). The VIA-IS was administered to Norwegian cadets prior to departing on a ten-week mission involving physically and mentally challenging tasks and a lengthy separation from family and friends, assessing their individual character strengths. Matching the individual strengths to their corresponding moral virtues (i.e., wisdom and knowledge, courage, humanity, justice, temperance, and transcendence)[13], it was found that the virtues had a marginal influence ($p < .15$) on self and peer ratings of productivity, self-confidence, and leadership behavior. The two studies of Matthews, Brazil, and Erwin (2009) and Matthews (2009) looked at character strengths and performance of soldiers deployed in actual combat conditions. They surveyed Army officers deployed in combat settings or those who recently returned from deployment, to investigate which strengths are most important to these combat leaders. The strengths consistently most frequently mentioned as relevant to military leaders in combat were bravery, citizenship, persistence, social intelligence, integrity, capacity to love, and judgment. Furthermore, the role of values, virtues, and character strengths in coping and resilience is of particular relevance to the military (Casey, 2011). The establishment of the Comprehensive Soldier Fitness program (Cornum, Matthews, & Seligman, 2011) as positive-psychology-based interventions to increase psychological strength and positive performance in the US Army reinforced the notion that character plays a key role in adapting and performing in combat and shows that within the military context it has been recognized that values and virtues are critical for a successful military profession.

In brief, these studies indicate that positive characteristics of soldiers must be inevitably taken into account to reliably describe and predict what makes a good, adaptive, and successful soldier. These and other results clearly suggest that positive psychology-derived constructs may contribute significantly to our understanding of how to train and educate soldiers. It is important to learn how values, virtues, and character strengths may play a role for the success of soldiers experiencing extremely challenging training and combat situations. In other words, the three pillars of positive psychology – positive states, positive traits, and positive institutions – provide a framework for pursuing research and application of positive psychology principles to military psychology.

---

13 Although the VIA-IS is providing strengths-based measures, as part of the statistical analysis in this specific study the matching of the 24 individual character strengths to the six virtues was made exceptionally according to the VIA classification by Peterson and Seligman (2004).

In accordance with Matthews (2008), character strengths that are important in combat can differ from those vital to success in training or in administrative job within the military. Whereas military institutions like the US Army hold their main focus on operational military targets and missions, the Swiss Armed Forces is focusing on training for operational readiness. In spite of this difference in missions, the Swiss Armed Forces is likewise an ideal setting for applying the principles of positive psychology. The first study in confirming the value of applying positive psychology to research within the Swiss Armed Forces was conducted by Eggimann and Schneider (2008). They studied character strengths and virtues of Swiss career officers and found hope, curiosity, vitality, bravery, integrity, and self-regulation to be significantly related to higher work satisfaction (cf. Proyer et al., 2012).

3 Research on values and virtues

Within numerous social science domains (e.g., sociology, political sciences, ethics), values and virtues play an important role, frequently the prime one (e.g., for explaining the circumstances of a value shift, voting behavior, moral judgment). It is recognized that values and virtues attract increasing interest and exhibit a large diversity of influence, and deserve significant research focus (Trommsdorff, 1996). The concepts of values and virtues, however, have been variously interpreted and broadly explored, both within theoretical and practical contexts. It was predominantly the philosophical viewpoint that initially stimulated the core discussions on the subject of values and virtues as psychological concepts (Urban, 1907; Münsterberg, 1908).

The following sections therefore reference the precursors in philosophy, describing the subject of values and virtues and its conceptual difference. Additionally, the corresponding theoretical approaches within the psychological literature are addressed. A special focus will be given to the relevance of values and virtues within personality psychology as well as positive psychology. Furthermore, areas of research requiring further attention will be highlighted within the following part.

3.1 Values and virtues in philosophy

Traditionally, it was the domain of philosophy to explore the nature and meaning of morally good characteristics such as values and virtues (Morales-Vives et al., 2014). Numerous moral philosophers and religious thinkers throughout history have been recognized for their interpretations of values and virtues. From a historical viewpoint, it has been an ongoing attempt to clarify, conceptualize, and formulate what is or should have value in individual and social life. The very first Greek philosophers asked "What is the good of a person?" This further inspired thinkers like Plato, Socrates, and Aristotle to examine and enumerate values and virtues as positive characteristics of individuals. Accordingly, it is crucial to include the historical perspective as part of ongoing research. The following provides an overview on how historical predecessors addressed and defined value and virtue.

It is worth mentioning that the concepts of values and virtues have been subject to separate theories in early areas of philosophers (Russi, 2009). Virtues were considered as the behavior that enable a person to align with a very important personal value and to have it preserved. A value reflects the anticipated goal, while virtues transform into action and behavior that allows a person to achieve the goal. Accordingly, the great philosophers derived their virtues from the presumptions that they made of the most important values. Correspondingly, the philosophers have either relied on the subject of virtue and emphasized more the behavior component or focused on values as the goal of behavior. For this reason, the following historical summary addresses both values and virtues in each individual, and in most of the situations the philosopher is dealing with either one of the concepts (Russi, 2009).

The sophists in ancient Greek were very competent professional teachers and intellectuals. They turned against an unconditional acceptance of moral values and inherently derived laws and norms. Protagoras of Abdera (490–420 BCE) was the most prominent member of the sophistic movement and believed "man is the measure of all things." He subscribed to the subjective view that values, virtues, and norms were the result of human agreement and could be rationally discussed and challenged (cf. Lee, 2005).

Plato (427–347 BCE), in reaction to this, was of a strongly different opinion: He was in search of the good in himself and anchored it as the highest idea within the absolute framework of ideas of the ideal human being. He was underlining the viewpoint that the good is the foundation of all values, but even more so of all being at all. Given this perspective, Plato also derived the four cardinal virtues. In the *Republic* as his magnum opus on the ideal human society, he defined the first major virtue catalog of the West: wisdom (*sophia*), courage (*andreia*), self-restraint (*sophrosyne*), and justice (*dikaiosyne*) (Plato, trans. 1968, as cited in Peterson & Seligman, 2004).

Aristotle (384–322 BCE) was Plato's most accomplished student. In his *Nicomachean Ethics* (Aristotle, trans. 2000) he tried to determine the highest good. In raising this question "What is the good?" Aristotle was not looking for a list of items that are good. Instead, his search was for the highest good. Consequently, he shifted the concept of good into the practical way of perspective: All people strive for happiness (*eudaimonia*), but everyone understands something different. In order to determine the greatest possible happiness for man, he has to acknowledge himself as a human species, which distinguishes him from animal in having reason. Consequently, the highest good, which is possible for human beings, must lie in practicing reason. Accordingly, human happiness is best achieved through a life of contemplation, which is the ultimate goal of human action and desire. The key question lies in how human beings can reach this objective. To this end, Aristotle developed a theory on virtues, which is called "virtue ethics" (Aristotle, trans. 1984). As part of his reasoning, Aristotle stated that virtuous behavior is a social practice exercised by a citizen of an ideal city. He had the understanding that virtue is an acquired skill learned through trial and error. Therefore, virtue is not inherent, but must be acquired in a theoretical and practical learning process. Whatever way virtue is learned, most important is the knowledge on the right amount and mean of a virtue. Related to this understanding Aristotle characterized virtues developing a doctrine of means: If a person encounters a situation and, basing the decision on reason, experience, and content, he or she chooses a course of action from between two extremes of dispositions, those of deficiency and excess. The mean between these two is virtue. Consequently, Aristotle conceived virtues as the desirable mean states between vices of deficiency and vices of excess. For instance, courage is the mean between cowardliness (vice of deficiency) and recklessness (vice of excess). Similarly, humility would be a virtue between the deficiency of shyness and the excess of shamelessness (Aristotle, trans. 2000, book II, chapter VIII). Central to this conceptual understanding is that individuals make a deliberate, rational choice to act in a manner that lies between these two extremes and is thus considered virtuous (Mintz, 1996). In fact, Aristotle assumed that good judgment held the greatest importance in ethics. Accordingly, the ability to carefully consider how a virtuous person would act when facing an ethical dilemma is key to developing a virtuous character (Cameron, 2011; Nybert, 2007; Solomon, 1992). As mentioned above, Aristotle's list of virtues includes the original four cardinal virtues. However, he added a number of other virtues, such as generosity, wit, friendliness, truthfulness, magnificence, and greatness of soul (Peterson & Seligman, 2004).

The philosopher Epicur (341–270 BCE) has declared joy to be the highest value for men and is considered the inventor of hedonism. Blessed is a life when it is free from physical pain (*aponia*) and free from confusion of the soul (that state he called *ataraxia*). Epicur did not enumerate a list of virtues, but he recommended staying out of many areas that required virtue (e.g., politics or marriage). Specifically, friendship between individuals was of great value to him (Krobath, 2009).

The Stoics further reinforced the value of ataraxia to apathy (*apatheia*): The path to happiness for the human being is found by not allowing ourselves to be controlled by our desire for pleasure or our fear of pain (Krobath, 2009). From Aristotle, the Stoics accepted that happiness was the highest value for human beings. This happiness, they taught, could be achieved through virtue, self-education, and self-control.

As a Christian theologian of the middle age, Aquinas (1224–1274) rejected Aristotle's additions to Plato and added three theological virtues proposed by Saint Paul: faith, hope, and love (Aquinas, trans. 1989, as cited in Peterson & Seligman, 2004). Aquinas argued for a hierarchical organization of virtues and defined the seven heavenly virtues: wisdom, courage, self-restraint, justice, faith, hope, and love. Within these seven heavenly virtues, Aquinas specified what Peterson and Seligman (2004) later defined as the six core virtues, describing transcendence with the virtues faith and hope, and humanity with the virtue of love.

In response to the stifled values and norms of the Middle Ages, the Renaissance, with its retrospectual revival of the classical thinking of ancient times and its ideal of the human personality, brought about a cultural crisis and the shattering of moral orders. Reformation and counterreformation stabilized values. With Luther (1483–1546) and Calvin (1509–1564), a canon of virtues and values was developed, that, until recently, was predominately accepted as a standard: orderliness, cleanliness, thriftiness, punctuality, industriousness, and diligence (cf. Bollnow, 1958).

A radical change towards modern value thinking then brought the Enlightenment with its appeal to reason. For Kant (1724–1804), the center of virtue is morality and the human end purpose of creation, and thus its supreme value. As a being endowed with reason, a person has the duty to strive after the good, according to what is accepted as morally correct. When all human beings fulfill their ethical duties, they generate values that benefit everyone. The main values that emerged in the course of the Enlightenment period are still valid in the broader sense, specifically humanity and human dignity, tolerance, individual freedom, and equality (Kant, 1785, as cited in Krobath, 2009).

Around the turn of the 19th century, the concept of value by Lotze (1817–1881) became a fundamental category of philosophy. Value theory is concerned with two fundamental questions: 1) What is value in itself? and 2) What are the different forms of values? Accordingly, two main groups can be distinguished: objectivists and subjectivists. The objectivists claim that things and actions can be evaluated because they have an absolute value (as cited in Krobath, 2009). Value judgments are therefore to be designated as true or false in the same way as descriptive judgments. The subjectivists claim that values are nothing other than projections of subjective feelings and attitudes. When you evaluate a thing or action, you do not say anything about it in itself, but instead express a subjective feeling or personal attitude. Moral values can therefore be neither true nor false (Hügli & Lübcke, 2003).

In the area of the debate about the objectivist view, axiology is described as a strict theory of values. Here certain axioms had been set up. Brentano (1838–1917) was the first to develop a classical theory of intrinsic value, which he attempted to base upon the philosophical psychology. In his essay "The origin of the knowledge of right and wrong" (1889), he presented fundamental considerations on ethical values. He thus founded the idea of "descriptive psychology" and shaped psychology as an exact science (Baumgartner & Reimherr, 2006). Specifically, he researched the criteria that relate to the question of law, custom, and order. Brentano was inspired by Aristotle's method to decipher and decode the essence of things by analyzing their simplest components and their structural contexts. His reasoning is based on the assumption that even for ethical problems, rational criteria of assessment and a justifiable ranking of values can be found (Brentano, 1889, as cited in Chisholm, 1986). By carrying out the possibility of such gradation, he presents a concept of value ethics. In doing so, he distinguished three fundamental categories of consciousness: imagining, judging (whether right or wrong), and emotional states (emotions that are good or bad, in the form of loving or hating). He thus presented a theory of values of the inner world and sought a way to place judgments of subjective feeling on a rational basis. Brentano recognized the origin of the concepts of truth or falsity, or of the good and the bad. "True" is something when the recognition referenced to it is appropriate, and "good" is something when the related love is correct. He also presupposed that universal and immutable moral laws exist for all humans. In addition, the following axioms were defined: the existence of a positive value is itself a positive value; the existence of a negative value is itself a negative value; the nonexistence of a negative value is a positive value, the nonexistence of a positive value is a negative value; the same value cannot be positive and negative at the same time; and it is impossible to maintain the same value for positive and negative (Brentano, 1889, as cited in Chisholm, 1986). According to the theory of Brentano, judgments based on reasoned criteria can be verified or falsified. Brentano was thus the first to create a foundation for the assessment of intrinsic values to relate to the subjective individual view, and at the same time to maintain an objective mindset.

The primary lessons learned from this historical excursion into the field of philosophy are the following:

- Values and virtues had been understood early in history as separate concepts;
- The concern of the greatest philosophers since antiquity has been with the question of "What are the most important values and virtues for humans?"
- There is variability across history, cultures, and intellectual tradition in terms of what values and virtues are worth striving for, but convergence can be found in the usually hierarchical listing and organizing of values and virtues;
- Plato stated that the good is the foundation of all values and derived the four cardinal virtues of wisdom, courage, self-restraint, and justice;
- Aristotle significantly shaped the notion of a virtue and assumed that virtues are moral qualities attributable to individual reasoning behavior, formulating a list of virtues including the original four cardinal virtues and a number of other virtues;
- Aquinas added three theological virtues (faith, hope, and love) to the original four cardinal virtues, resulting in a categorization very similar to Peterson and Seligman's (2004) understanding of six core virtues; and
- Bretano developed a classical theory of intrinsic value and founded the rational basis for judgments of subjective feeling on values.

Interestingly, many of the central ideas of these philosophical thinkers later reappeared and contributed to psychological approaches towards values and virtues. Therefore, it is essential to consider the historic perspective on value and virtues, when interpreting current and future research findings. Subsequently, in the description of the theories on universal values and virtues below, the two concepts are discussed in separate chapters.

3.2 Universal values: Definitions and theories

Scholl-Schaaf (1975) defined three types of values: (a) value defined as a guiding principle, (b) value as a norm, and (c) value as a goal. The first definition, value as a guiding principle, is the basis of this research; however, all three definitions are given consideration. A value refers to the individual importance and relevance of a particular subject: "It is of great value to me," meaning "It is of importance to me and I will stand up for it."

According to Bilsky (2005), the attempt to agree on a unified definition of values has been unsuccessful. This implies that comparing and interpreting research results has its limitation in terms of precision and reproducibility, unless a particular value has a common lexical definition. Nevertheless, there are predominant factors of commonality when reference is made to values. Hitlin and Piliavin (2004, p. 362) described the domain of values as an "internal moral compass." A more general definition of value is given by the lexicon of Dorsch (2016, 18th edition, p. 1790):

> With reference to individual values by Kluckhohn (1951), values are defined as explicit or implicit conceptions of the desirable, both in the context of an individual and a group, impacting the choice between available types, means and goals of actions. This often criticized conceptual formulation (Graumann & Willig, 1983) has not been substituted in literature through a more agreeable definition (Rohan, 2000).

In alignment with the definition above, there is consensus within research, that in the context of values "there is a relatively limited number of concepts or descriptions, which correspond to desirable behaviors or goals (or mental states)" (Bilsky, 2005, p. 300). Moreover, it is assumed that values are valid across situations, imposing guidance towards choice and appraisal of behaviors and circumstances. In line with this understanding, Rokeach (1973b, p. 5) defined values as "enduring beliefs that a specific mode of conduct or end-state of existence is personally or socially preferable to an opposite or converse mode of conduct or end-state of existence." He was interested in a full set of values as "guiding principles" (p. 358) to describe an individual view, and implemented two distinct lists of 18 instrumental values (describing modes of conduct as forms of behavior) and 18 terminal values (describing end-states of existence as lifetime goals). Similarly, Schwartz (1992, 1994, 1999) focused on the motivational power of values and defined them as desirable goals that vary in importance across situations and that guide the way social actors (e.g., individual persons such as military leaders) choose actions and evaluate people and events. He derived his definition of values from the understanding of Rokeach and conceived values likewise as "cognitive representations of desirable, abstract, trans-situational goals that serve as guiding principles in people's lives" (cited according to Corr and Matthews, 2009, p. 593). Unlike Hofstede (1980) or Schwartz (1994, 1999) who were interested in values as they manifested themselves at a sociocultural level, Rokeach studied values as interindividual differences. As an overview, Schwartz and Bilsky (1987) concluded on five formal features of values, which are usually addressed in definitions: "According to the literature, values are (i) concepts or beliefs, (ii) about desirable end states or behaviors, (iii) that transcend specific situations, (iv) guide selection or evaluation of behavior and events, and (v) are ordered by relative importance" (p. 551).

The above definitions of values illustrate that the application of the concept of values implies individual, social, and organizational aspects. In situations where individual values take preference, the corresponding research is assigned to personality psychology, while research on social values align with social psychology. In concrete, personality psychology concentrates on:

- measuring personal values as interindividual differences;
- distinguishing a set of core values, to categorize them and to investigate their structure; and
- reaching a conclusion as to how values relate to basic traits of personality (Bilsky, 2005).

Considering the publications on values of the last decades, one recognizes that the expanding interest in psychological value research correlates with the increase in social psychological research. Within the context of personality-psychology, the actual efforts in research were diminishing. An analysis of the research activities on ISI Web of Science (December 5, 2017) showed 8,433 records on *values* within the psychological field. More than half of these manuscripts have reference to social psychology and among the ten most cited (ISI Web of Science, "citation report values," 2017) are six articles which address the scope of social psychology. In accordance with Bilsky (2005), the constraint is with research, which aim at a theoretical-based integration of available research data. As proposed by Bilsky and Schwartz (1994), as well as by Roccas, Sagiv, Schwartz, and Knafo (2002), the future demands a substantial degree of coordinated research. This is confirmed by De Raad and Van Oudenhoven (2008), who declared: "The study of values had an almost equally long history as that of traits, but catalogues of values that claim full and integrated coverage of the field are hard to find" (p. 82).

The present thesis had a personality psychological focus and studies values as they pertain to individual military persons. Its aim is to make an integrative contribution to the existing approaches to values, establishing a catalog of values with specific reference to military psychology. In alignment with previous studies on interpersonal differences in values, in the present thesis a value was understood to be "a relatively enduring characteristic of individuals that reflect what is important to them and that guides them in their behaviors and decisions" (De Raad & Van Oudenhoven, 2008, p. 85–86).

3.2.1 Two main strains of value research

The following section reviews the progress in psychological research on values, including the appropriate theoretical background and understanding.

The psychological systems of values are focused on the question of how many universal values can be distinguished. This particular subject has been investigated within two domains of research, with little connection to each other, referring back to (1) Spranger (1928) as well as Vernon and Allport (1931), and to (2) Rokeach (1973b) and Schwartz (1992, 1999). It is worth mentioning that the empirical value research played a lesser role within psychology during the first half of the 20th century and the first strain of research pursuit (Bilsky, 2005). The situation only changed towards the end of the 1960s in the second strain of research, in reference to Milton Rokeach, who enjoyed wide acceptance within the domain of psychology. Likewise, Schwartz (1992) aligned with Rokeach (1973b) value theory, in an attempt to explore a universal theoretical structure of values. The following sections describe more specifically the two separate research strains on values.

1a) Types of Men by Spranger
Eduard Spranger was a German philosopher, pedagogist, and psychologist, widely acknowledged for his contribution in establishing pedagogy as a distinct academic discipline. Spranger's most influential contribution to personality theory was his book *Lebensformen* ([*Types of Men*]; Spranger, 1928).

Spranger (1928) held the opinion that the human personality is best understood by assessing the corresponding values. To facilitate the process of exploring knowledge, Spranger conceptualized the six "ideal types of individuality" based on rational thinking. This perspective accounts for the theoretical, the economic, the aesthetic, the social, the political, and the religious type. According to Spranger (1928), the various personality types and their corresponding values are characterized as follows (information retrieved from Krobath, 2009):

- Theoretical type: a passion to discover, systemize, and analyze; corresponding values include discovery of truth, discovery of rules, and search for knowledge.
- Economic/Utilitarian type: a passion to gain a return on all investments involving time, money, and resources; corresponding values include work, production, security, and wealth.
- Aesthetic type: a passion to experience impressions of the world and achieve harmony in life; corresponding values include individual expression, fantasy, beauty, harmony, and grace.
- Social type: a passion to invest myself, including my time and resources, into helping others achieve their potential; corresponding values include love, empathy, and loyalty.
- Political/Power type: A passion to achieve position and to use that position to affect and influence others; corresponding values include power, vitality, perseverance, self-realization, and influence.

– Religious/Traditional type: A passion to seek out and pursue the highest meaning in life, in the divine or the ideal, and achieve a system for living; corresponding values include unity and relativity of human existence.

According to Spranger (1928), this concept is a logical consequence of the assumption, "that each individual structure has a dominating value orientation" (p. 114). Spranger implied that no human personality was bound to fall exclusively into the scope of one of the six "ideal types of individuality."

1b) "Study of Values" by Allport and Vernon

The studies by the psychologists Allport und Vernon in the 1930s initiated a first step towards an empirical research on values. Furthermore, Allport and Vernon (1931) established the "Study of Values" scale (SOV). This was the most generally used instrument in psychology for many years, based on Spranger's rationally created model of the six "Types of Men" (cf. Lurie, 1937). The SOV was revised in 1951 (content-wise) and in 1960 (formally). Thereafter the "Study of Values" received strong international acceptance, although it was criticized for perceived "conservative-educational" items and for confounding effects in values and interests (Graumann & Willig, 1983). Specific information on this measurement instrument is provided in section 3.2.2.

1c) Psychology of Politics by Eysenck

In his book *Psychology of Politics*, personality psychologist Eysenck (1954) referred to values and interpreted the six value types described by Spranger. By analyzing various statements on the subject of social attitude, he identified two orthogonal personality factors which he named "tough-tender Mindedness" (T-factor) and "Radicalism-Conservatism" (R-factor) (cf. Ray, 1973). He assumed that each value type by Spranger (1928) is dominated by a different value attitude as captured by the T- and R-factor. Furthermore, Eysenck described various evaluation studies about the SOV and drew conclusions in his final part: "The evidence is fairly conclusive that the values as measured by the Allport-Vernon Scale are closely related to interest patterns and that these interest patterns show a structure well in line with our T-factor" (p. 169). However, Eysenck (1954) did not claim that the T-factor is a sufficient dimension of covering social attitudes. Accordingly, Eysenck searched for personality psychological explanations of why tough-mindedness (among the T-factor) goes beyond the continuum of left-right orientation (e.g., Stalin vs. Hitler). By investigating empirical data he concluded on the schematic solution that the fascist-right mindset is identified as a tough-minded conservative position and the communist-left mindset as a tough-minded radical position. He reasoned that the R-factor fundamentally includes social attitudes, and the T-factor can be seen as a projection on to the attitude level of a personality variable of Extraversion. By drawing a relationship to values, Eysenck succeeded in a first attempt to conceptually combine the two personality factors tender-mindedness vs. tough-mindedness (T-factor) as well as Radicalism vs. Conservatism (R-factor) with the value types of Spranger.

1d) Other systems of values

In the 1940s and 1950s Morris (1956) developed a different, equally well-respected, research approach. As a theoretical basis, he referred to the assumption of the three components of dependence, dominance, and detachment (named as Dionysian, Promethean, and Buddhist), which were considered fundamental to human personality (cf. Schlöder, 1993). He defined the concept of seven "Lebenswege" ("ways of life"), each of which he assigned to one of the three components: the Buddhist path of detachment of desire, the Dionysian path of abandonment to primitive impulses, the Promethean path of creative reconstruction, the Apollonian path of rational moderation, the Christian path of sympathetic love, the Mohammedan path of the holy war, and the path of generalized detachment-attachment. From an empirical perspective, his initial theoretical approach was too constraining. Therefore, Morris and Jones (1955) extended the theory to thirteen "ways of life," which were assessed by participants on the basis of a combined rating and ranking scale (Braithwaite & Scott, 1991). Richards (1966) factor-analyzed ratings from college freshmen in 31 institutions of higher education on 35 items pertaining to the students' goals. The factor-analysis was performed separately for males and females. The two groups were found to share seven factors, which according to Richards (1966) assess many of the same factors of the SOV by Allport and Vernon (1931).

2a) The system of values by Rokeach

As another theory, Rokeach (1973b) provided with his publication "The Nature of Human Values," an influential theory concerning the way values are understood. His assumption was that i) the number of values held by an individual person is relatively small, ii) all human beings have the same values differing in their extent, and iii) values are organized into value systems. Accordingly, he was interested in a full set of values to describe an individual view. Rokeach implemented two distinctive lists of 18 instrumental values (describing modes of conduct as forms of behavior) and 18 terminal values (describing end-states of existence as lifetime goals).

The Rokeach Value Survey (RVS; Rokeach, 1973a) was widely used as a comprehensive psychological instrument to assess individual values (e.g., Braithwaite & Law, 1985; Feather, 1986; Rokeach, 1973b, 1979). The 18 terminal values, covered by nouns, and 18 instrumental values, covered by adjectives, are sorted into an individual ranking according to the degree the value is desired for one's self and for others (referring to Tab. 2).

Table 2 identifies the two lists of terminal and instrumental values of the RVS. The instrumental values correlate well with concrete modes of conduct as norms of behavior[14], while terminal values are positioned on a higher abstraction level. Terminal values can be attained by varying different modes of conduct. For example, social recognition (terminal value) can be accomplished through instrumental values such as polite, capable, cheerful, forgiving, responsible or helpful (Asendorpf, 2004). Factor-analysis based on rank-orders of the two lists of values showed six factors (Rokeach, 1974; Rokeach & Ball-Rokeach, 1989), called (1) immediate vs. delayed gratification, (2) competence vs. religious morality, (3) self-constriction vs. self-expansion, (4) social vs. personal orientation, (5) societal vs. family security, and (6) respect vs. love. Although Rokeach's theory was aimed at differentiating between instrumental and terminal values, he assumed that instrumental and terminal values could be further specified, dependent on whether they relate to individual wellbeing or to the wellbeing of others. The instrumental values with individual focus he called "competence values" (e.g., to be ambitious, intellectual, or independent) while those instrumental values with the wellbeing-focus on others he called "moral values" (e.g., to be helpful, forgiving, or polite). Accordingly, terminal values with self-focus were called "personal values" (e.g., self-respect, comfortable life, or freedom), and those with the focus on others were addressed as "social values" (e.g., equality, national security, or a world at peace). Ultimately, Rokeach's value system is based on the assumption that the differentiation between self-focus and other-focus is of high significance. Rokeach (1973b) concluded: "Values are the joint results of sociological as well as psychological forces acting upon the individual" (p. 29). With this perspective in mind, values represent the personal and the social preferences of individual needs and social norms. The model by Rokeach received broad acceptance within the psychology discipline, mainly justified by

---

14 As further outlined in section 3.3 on virtues, the conceptualization of instrumental values by Rokeach (1973b) resembles the understanding of virtues as a morally good trait, enabling a person *to behave* and *act* in accordance with his or her personal values (De Raad & Van Oudenhoven, 2011).

Tab. 2: *Values of the RVS (Rokeach, 1973a)*

Terminal values (nouns)	Instrumental values (adjectives)
comfortable life	ambitious
exciting life	broadminded
a sense of accomplishment	capable
a world at peace	cheerful
equality	clean
family security	courageous
freedom	forgiving
happiness	helpful
inner harmony	honest
mature love	imaginative
national security	independent
pleasure	intellectual
salvation (belief in God)	logical
self-respect	obedient
social recognition	polite
true friendship	responsible
wisdom	self-controlled

the RVS being an economical and widely applicable instrument. Nevertheless, it was not based on a coherent theory of values, instead being the result of a series of unconnected, predominantly plausible assumptions (Bilsky, 2005). The inadequate theoretical justification of Rokeach's research approach led to repeated attempts to validate the inherent structure of the RVS.

2b) The system of values by Schwartz

In recent years, the conceptual framework of Schwartz (1992, 1994) and Schwartz and Bilsky (1990) has strongly influenced the research on values. From the outset, Schwartz was interested in developing a structural theory of human values, which can take cultural-specific and cross-cultural aspects into equal account. His approach was designed to summarize the multitude of unrelated individual values, which differ in their motivational content. By means of a facet-theoretical approach, he initially identified seven motivational domains of values and ultimately increased it to ten domains, representing a universal structure of the ten human values of universalism, benevolence, conformity, tradition, security, power, achievement, hedonism, stimulation, and self-direction.

Tab. 3: *Definitions of the ten value types by Schwartz (1992), cited according to Mohler and Wohn (2005)*

Value type	Definition	Covering values
1 Universalism	understanding, appreciation, tolerance, and protection for the welfare of all people and for nature	broadmindedness, social justice, equality, world at peace, world of beauty, unity with nature, wisdom, protection of the environment
2 Benevolence	preserving and enhancing the welfare of those with whom one is in frequent personal contact (the "in-group")	helpfulness, responsibility, forgiveness, honesty, loyalty, mature love, true friendship, meaning in life, sense of belonging
3 Conformity	restraint of actions, inclinations, and impulses likely to upset or harm others and violate social expectations or norms	obedience, self-discipline, politeness, honoring parents and elders, loyalty, responsibility
4 Tradition	respect, commitment, and acceptance of the customs and ideas that one's culture or religion provides	respect for tradition, humility, modesty, spiritual life, respecting my "portion" in life
5 Security	safety, harmony, and stability of society, of relationships, and of self	social order, family and national security, reciprocation of favors, health
6 Power	social status and prestige, control or dominance over people and resources	authority, wealth, social power
7 Achievement	personal success through demonstrating competence according to social standards	ambition, influence, success, intelligence
8 Hedonism	pleasure or sensuous gratification for oneself	pleasure, enjoying life, self-indulgent
9 Stimulation	excitement, novelty, and challenge in life	a varied, exciting, and daring life
10 Self-Direction	independent thought and action	creativity, freedom, independency, choosing own goals, curiosity, self-respect

Initially, Schwartz and Bilsky (1987, 1990) analyzed the RVS of Rokeach in seven different cultures with partially varying methods. The similarity of the individual rankings was measured within each culture by correlations. Subsequently, values were projected on to a two-dimensional space of similarities via nonmetric multidimensional scaling. With this method, regions of homogeneous values emerged, which can be illustrated geometrically on a surface. This method was used to show that the seven investigated cultures were able to distinguish between seven homogeneous areas of values: hedonism, security, achievement, self-direction, restrictive-conformity, benevolence, and universalism. Through additional intercultural empirical studies, Schwartz (1992) was able to prove the existence of ten different types of values as well as their linkage to two underlining dimensions in more than 40 countries. Based on the intercultural universality of the types of values, he derived the motivational basis of values. In his opinion, they are responsive to three universal needs of human existence: biological needs, the need for coordinated social interaction, and the need for the proper functioning and the survival of the groups. Table 3 shows the ten homogeneous value ranges as well as the corresponding definitions.

As Table 3 shows, Schwartz (1992) assumed that values could be subdivided into ten motivational values. Each value category represents a specific motivational goal.

A further assumption of Schwartz (1992, 2006) refers to the reciprocal relations among the types of values. Actions that are carried out under the goal of a type of value have certain psychological, practical, and social consequences. These are consistent with the consequences of actions under pursuing the goal of other value types or not. The ten types of values form a continuum, which Schwartz typically presents as a circular structure (Fig. 2).

As Fig. 2 illustrates, Schwartz' theory contains a structural model that allows differentiated statements on the relationships between the ten different types of values formulated by Schwartz (1992, 2006). Within this circumplex of values, incompatible values lie opposite one another, complementary values supporting similar goals are to be found close together. In a next step, Schwartz identified the organization of the ten types of values on two bipolar dimensions. As Fig. 2 illustrates, the compatibilities between the ten types of values result in a circular structure that is based on two orthogonal dimensions. These two dimensions are formed by four higher-order types of values, known as standard types, and were described by Schwartz as "Self-transcendence" vs. "Self-enhancement" and "Openness to change" vs. "Conservation." The first dimension is stretched by two types of values, "Openness to change" and "Conservation." This dimension contains values that, on the one hand, emphasize independent thinking, actions, and varied opportunities, and on the other hand, emphasize obedient self-restraint, preservation of security, and traditional action. In this dimension, the values of hedonism (8), stimulation (9), and self-direction (10) (~liberal) are opposed to the types of security (5), conformity (3), and tradition (4) (~conservative). The second dimension is formed

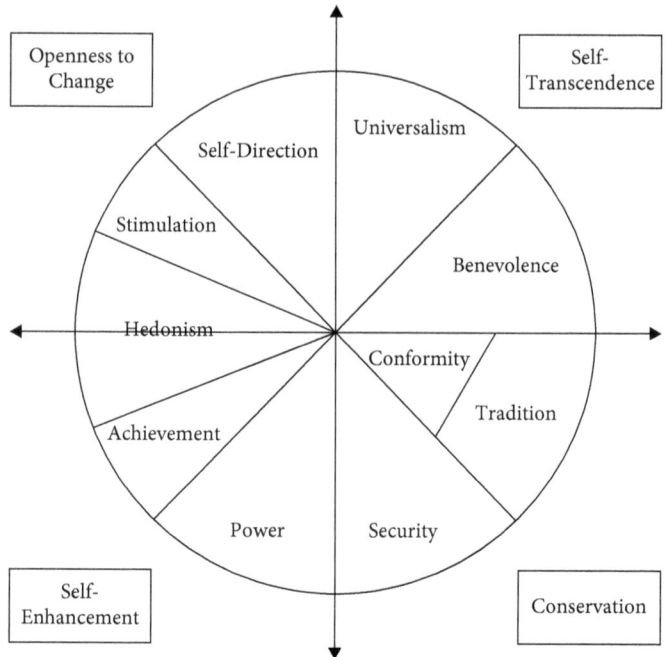

Fig. 2: Theoretical model of relations among the ten motivational types of value as proposed by Schwartz (2006) (illustration drawing from Schwartz, 2006)

by the values of higher order, Self-enhancement vs. Self-transcendence. Here, values that accentuate the acceptance of others as equal individuals and the benevolence of others are opposed to those values that represent a striving for one's own success and dominance over others. The quadrant of achievement (7) and power (6) (~ hard) as one of the pole of the superior dimension stands opposite to the quadrant of universalism (1) and benevolence (2) (~ soft) as the other pole. Schwartz' approach has been examined in a multitude of international studies and has largely been confirmed in terms of its basic assumptions (content and structure of values; cf. Schwartz & Sagiv, 1995).

2c) One-dimensional view of human values with individualism and collectivism by Hofstede
One of the most prominent researchers on values in the last quarter of the 20th century is Geert Hofstede. Hofstede (1980, 2001) concentrated on national-level patterns of values, assuming that values constitute a stable portion of national culture. His approach to the value domain is described as follows: "The core

of culture (...) is formed by values. Values are broad tendencies to prefer certain states of affairs over others" (Hofstede, 1997, p. 8). In this same study, he used a workplace questionnaire to evaluate the opinions of employees from a multinational corporation (with employees in 39 countries), on managerial decision-making styles and needs, and the importance of different values. From his data, he derived five statistically independent values dimensions: power distance (acceptance of inequality), uncertainty avoidance (the way societies deal with the uncertainty of the future), individualism vs. collectivism (the individual's dependence on the group), and masculinity vs. femininity (the social implication of being a man or a woman). Hofstede never focused on explaining individual behavior and warned against applying his findings on the level of individuals. Moreover, his approach has been variously criticized, especially regarding the low representativity of the sample and the poor validity of the selected items he used for his questionnaire (Hansen, 2000). In spite of these deficiencies, Hofstede introduced a central view that people and societies seem to vary in terms of an emphasis on the importance of individuality on one hand, and on the importance of being part of a larger social community on the other hand. This one-dimensional individual-collectivism distinction was named the "deep structure" of cultural differences (Greenfield, 2000), and has been applied to describe differences in values, attitudes, and behaviors between national groups and cultures (Oyserman, Coon, & Kemmelmeier, 2002). Collectivism suggests that individuals may subordinate their personal goals to the goals of a collective (e.g., family, co-workers, or company). In individualistic societies the main unit is the individual, the cohesion is loose, and everyone looks after himself or herself (Hofstede, 1997). A similar concept in personality research is the two-dimensional concept introduced by Bakan (1966), and adopted by Wiggins (1991), of agency and communion. The distinction between these two independent dimensions has been shown to be valid in capturing individual differences of interpersonal traits (e.g., DeYoung, 2006; Digman, 1997; Saucier et al., 2014).

This perspective in this conceptualization of agency-communion and individualism-collectivism points to the fact that different cultures may stress different kinds of values, which could also shape different structure of values. Therefore, it is important to study the structure of values in different cultural environments such as, within the framework of this thesis, in the specific culture of the Swiss Armed Forces as a distinct military organization.

At the end of this outline on the different strains of value research, it must be emphasized that Spranger (1928), Allport and Vernon (1931), Eysenck (1954), and Rokeach (1973b) primarily focused on the differences in values between individuals. Schwartz was a social psychologist and interested in differences between culture, with establishing a framework of values which captures universal structure of values (Schwartz, 1992, 1994). As De Raad and Hendriks (1997) mentioned, the theoretical background and the methodological approaches in developing the value facets have an influence on the development of the corresponding measurement instruments. Therefore, in the following section, a selection of existing self-report instruments on values and their approach in development will be presented.

3.2.2 Measurement instruments

This review focuses on five well-established instruments assessing individual value preferences: the *Study of Values* (SOV; Kopelman, Rovenpor, & Guan, 2003), the *Rokeach Value Survey* (RVS; Rokeach, 1973a), the *Schwartz Value Survey* (SVS; Schwartz, 1992), the *Portrait Values Questionnaire* (PVQ; Schmidt, Bamberg, Davidov, Herrmann, & Schwartz, 2007), and the *Austrian Value Questionnaire* (AVQ; Renner et al., 2004). Although all four instruments measure values, they pertain to different theoretical backgrounds and therefore focus on different aspects of value definitions. An important methodological point to mention is the question of the type of scoring, specifically, using a ranking-format versus a rating-format. The ranking-format is a forced choice scale, in which respondents have the option of two or more desirable answers and are asked to choose the one that they most prefer (Baron, 1996). This is contrasted with instruments that use Likert-type scales in rating-formats: respondents are instructed to indicate the score (e.g., 1 to 5), which best represents the degree of agreement with a given statement (Carifio & Perla, 2007).

In specific, the ranking-format is linked to an ipsative measure which means that it is individually referenced rather than norm referenced (Gaylin, 1989). Accordingly, a ranking-format assesses the relative prominence of each of the presented values within the respondent's ipsative profile, not the absolute strength of each value. In accordance with Ball-Rokeach and Loges (1996), ipsative measures imply the real-world notion that values are often in competition with one another. Moreover, ipsative measures include that "the sum of the scores over the attributes for each individual equals a constant" (Cheung & Chan, 2002, p. 58) and is equal to each of the respondents. Ipsative scores differ from normative scores relating the assessment of relative instead of absolute values (Brown, 2010), with the result of a clear profile of values. However, the analysis of ipsative data is seen as problematic, because standard statistical analyses turn out to be biased (e.g., Clemans, 1966; Hicks, 1970). As a consequence, only intra-individual, but not interindividual comparisons are possible (e.g.,

Baron 1996; Closs, 1996). According to Van Eijnatten, Van der Ark, Holloway, Eijnatten, and Ark (2015), the use of ipsative measures is mostly applied in research about organizational values, because ipsative measures may lower the risk of social desirability response bias (Meglino & Ravlin, 1998). Hicks (1970, p. 167) stated that: "each score for an individual is dependent on his own scores on other variables, but is independent of, and not comparable with, the scores of other individuals."

In comparison, the rating-format includes a direct rating, with the possibility that different values can be equally important. Rating-formats are therefore providing absolute values (Carifio & Perla, 2007). However, the rating-format can facilitate response sets, which is the tendency to score all value items as more important or as less important (e.g., acquiescence bias). Such a response set is most likely to show up on the undifferentiated values – the ones with a high universal social desirability. If data carry such a response set in a factor analysis, it tends to occupy the first factor, and it normally explains a larger share of the common variance than any other factor, which is content-related. There is consensus in the literature that this is an important point to be considered within analyzing and interpreting data (Meglino & Ravlin,1998).

The *Study of Values* (SOV) by Allport and Vernon (1931) was designed to measure Spranger's six value types (theoretical, economic, aesthetic, social, political, and religious), using a ranking format. The original SOV was revised in 1951 and again in 1960. However, outdated item content and wording occurred, as reported by Campbell, Jayawickreme, and Hanson (2015). To revive the instrument, Kopelman et al. (2003) published a fourth edition of the SOV (SOV4), in which they updated 15 of the original 45 items, maintaining the format and scoring of the questionnaire. According to Kopelman et al. (2003) the Cronbach alpha for the six value types ranged from α =.66 to α =.72, although the Cronbach alpha coefficient differed somewhat for the economic value domain (α =.64 for the original versus α =.72 for the revised version) and the political value domain (α =.61 for the original versus α =.55 for the revised version; Kopelman et al., 2003).

The *Rokeach Value Survey* (RVS) was developed for rank-order scaling of 36 values, including 18 terminal and 18 instrumental values. The task for participants in the survey is to arrange the 18 terminal values, followed by the 18 instrumental values, into an order "of importance to you, as guiding principles in your life." There have been many attempts to reduce the 18 instrumental values and 18 terminal values into a set of underlying factors, but without consistent success (Feather & Peay, 1975; Johnston, 1995). Hofstede and Bond (1984) demonstrated in a cross-cultural research the association of the RVS values with the Hofstede dimensions. Specifically, they revealed in correlation analysis that each of the dimensions can be distinctly identified by RVS measures which points to the universal nature of the RVS.

Based on the RVS, Schwartz (1992) developed the *Schwartz Value Survey* (SVS), containing 56 items, which includes two lists of value terms (specifically 30 terminal and 26 instrumental values). An explanatory phrase follows each item, and each item expresses the motivational goal underlying a single value (e.g., "equality [equal opportunity for all]" as a universalism item and "pleasure [gratification of desires]" as a hedonism item). Participants are asked to rate each item based on to what extent it is seen "as a guiding principle in my life," rated on a 9-point Likert-type scale from 1 (*not relevant as a guiding principle for me*) to 9 (*of supreme importance as a guiding principle for me*). Schwartz (1994) prefers the rating of value importance to ranking. To minimize response bias, he advises partialling out respondents' mean value ratings to ensure that SVS scores reflect within-person (ipsative) value priorities (Schwartz, 2012). Schwartz followed Rokeach's (1973b) idea that terminal values and instrumental values function differently. However, Schwartz (1992) suggested that this distinction has no substantive importance and participants could not differentiate between terminal and instrumental values. Accordingly, one item in the 56-item SVS was dropped and two others added in the revised 57-item version. Schwartz (1992, 2012) collected extensive data from 20 countries within cross-cultural studies and samples drawn from various occupational and age groups to validate the SVS. He reported stability coefficients across a six-week interval for all 10 SVS subscales that were .70 or higher. Test-retest reliability coefficients ranged from .70 (Self-Direction, Achievement) to .82 (Tradition). Pozzebon and Ashton (2009) reported Cronbach alphas ranging from α =.60 (Security) to α =.68 (Benevolence). Bardi and Schwartz (2003) found positive correlations between SVS scores and behavior rating value scales ranging from .30 (Benevolence) to .68 (Stimulation). Accordingly, the SVS has been widely validated in cross-national studies on differential value preferences. The approach by Schwartz (1992) goes beyond the application of the RVS, by including further categories of values. However, it cannot solve the main problem of the RVS, being the relatively random and intuitive selection of the values.

The *Portrait Values Questionnaire* (PVQ; Schmidt et al., 2007) is an alternative to the SVS developed in order to measure the ten basic values in samples of children (from age 11) and adults. The SVS had not proven suitable to such samples and the European Social Survey (ESS) gave preference to the PVQ to measure values. Equally important, to assess whether the values theory is valid independent of method required an alternative instrument. The PVQ includes short verbal portraits of 40 different people, gender-matched with the respondent.

Each portrait describes a person's goal or wish that points implicitly to the importance of a value (e.g., "Thinking up new ideas and being creative is important to him. He likes to do things in his own original way," describing a person for whom self-direction values are important). For each portrait, respondents are asked to indicate "how much they like this person" in a ranking-format, comparing the portrait to themselves rather than themselves to the portrait. All the value items showed near-equivalence of meaning across cultures in studies using multidimensional scaling (Schwartz, 2005). Across 14 samples from seven countries, Cronbach alpha reliabilities of the ten value types had an average of α =.68, ranging from α =.47 for Tradition to α =.80 for Achievement (Schwartz, 2005).

The *Austrian Value Questionnaire* (AVQ; Renner et al., 2004) was developed on the basis of the lexical approach to account for specific facets of values in German-speaking countries. Renner (2003b) made the distinction between instrumental values and terminal values, and identified value-describing nouns and adjectives from the German lexicon, administering the two lists to a large sample of German-speaking participants. The final version of the AVQ comprises 54 items which constitute five scales of value factors: Intellectualism, Harmony, Religiosity, Materialism, and Conservatism. Each item has to be rated on a 5-point scale how much the person supports or disapproves it as a guiding motive in his life. The AVQ has been validated in several studies and has proved to be a culture-specific measure, reliable, and valid instrument for measuring universal values among the German-speaking population (Renner, 2003a; Salem & Renner, 2004).

3.2.3 Values and personality

In their overview of literature and research on the relationship between values and traits of personality, Athota and O'Connor (2014) stated that due to the general theoretical understanding that values serve as guiding principles in a person's life, it is not surprising that "researchers have typically used values to help explain external criteria" (p. 51). For example, Chan and Drasgow (2001) as well as Clemmons and Fields (2011) found positive relations between values and motivation to lead; and Caprara, Vecchione, and Schwartz (2009) showed positive associations between values and social interaction. Athota and O'Connor (2014) concluded that many studies had focused on consequences of values but few had focused on related constructs of values such as personality.

In line with Dollinger, Leong, and Ulicni (1996) personality traits and values can be viewed as individual differences that are both cross-situationally and cross-temporally consistent. According to this study, it can be expected that values converge with personality traits and are similar psychological constructs. However, the empirical research on personality and values has taken independent avenues (Aluja & García, 2004). Accordingly, little research has been conducted to explore the relationship between traits of personality and values. Similar to the status within value theories, there are various conceptual approaches to personality (Carver & Scheier, 1992). Among these, the trait approach seems most appropriate for an integrative purpose: traits are "dimensions of individual differences in tendencies to show consistent patterns of thoughts, feelings and actions" (McCrae & Costa, 1990, p. 23). In line with this definition, Johnson (1997) described personality traits as relatively stable characteristics that determine how people think, feel, and act. According to Guilford (1959), personality is referring to the individual "unique pattern of traits" (p. 5). He differentiated between broad classes of traits that represent different aspects of personality: somatic traits (physiological), motivational traits (needs, interests, and attitudes), aptitudes, and temperaments. If we take Rokeach's (1973b) reformulation into account, stating that a person's character, "which is seen from a personality psychologist's standpoint as a cluster of fixed traits, can be reformulated from an internal phenomenological standpoint as a system of values" (p. 21), it seems most appropriate to link the motivational aspect of personality (i.e., people's needs, interests, and attitudes) with values as broad goals guiding behavior. As Bilsky and Schwartz (1994) outlined, "the motivational trait and the parallel goal vary together" (p. 166).

De Raad and Van Oudenhoven (2011) highlighted the differences between values and personality and specified that personality traits are characteristics that *dispose* a person to behavior, while values are characteristics that influence what a person finds *important* and that *guide* a person in behavior and decisions. In accordance, Roccas et al. (2002) pointed out that traits are enduring dispositions and values are enduring goals. However, values differ from specific goals because values are consistent across situations (Emmons, 1989; King, 1995; Roberts & Robins, 2001). Unlike the two other related constructs of needs and motives, values are basically desirable and must be represented in a cognitive manner, enabling people to communicate about them (Bilsky, 1998; McClelland, 1985).

As Roccas et al. (2002) pointed out, explicating the relationship of personality traits to values will deepen our understanding of both. The literature has specifically identified the Big Five factor of Agreeableness as morality-related personality traits (e.g., Strelau & Zawadzki, 1996) and concluded that they "may therefore be considered as the more typical value-laden character factors" (De Raad & Van Oudenhoven, 2008, p. 102). Substantial correlations between Agreeableness and values were found with Benevolence, Love,

Balance, Universalism, Security, and Conformity (positively) and with Status, Profit, and Comfort (negatively; Renner, 2003b; Roccas et al., 2002; De Raad & Van Oudenhoven, 2008). The study by Morales-Vives et al. (2012) confirmed that people with high levels in Agreeableness tend to value concepts such as Solidarity and Benevolence. Results from various studies (De Raad & Van Oudenhoven, 2008; Pozzebon & Ashton, 2009; Renner, 2003b) showed that people with higher levels of Conscientiousness tend to prefer values such as Responsibility, Organization, Achievement, Conservatism, Professionalism, Family, and Tradition. In the study by Renner (2003b), Conscientiousness correlated with the values of Conformity and Conservatism. Correlations to the other personality factors such as Extraversion and Openness to Experience were also found in the same referenced studies: people with higher levels in Extraversion tend to consider Hedonism, Happiness, and Social relationships as important factors in their lives. Openness to Experience was consistently found to correlate positively with values of Universalism, Self-Direction, and Stimulation and negatively with Security and Tradition. The trait factor Emotional Stability was consistently found not to significantly correlate with any value factors (Morales-Vives et al. 2012). The overall findings confirm that Big Five personality traits do correlate with values and do allow an indicative interpretation in accordance with the distinct meaning of a particular value descriptor. Moreover, correlations of values with facets of the five factors revealed nuances of the facets and clarified the meanings of the factors (Roccas et al., 2002). Again, no studies to date have investigated the specific relationship as it applies to military values and personality.

3.2.4 Values and motivation

According to Karp (2000), the study of values often addressed questions about what motivates human behavior, since values are seen as expressions of different motivational goals. As described above, the theories by Rokeach (1967) and Schwartz (1992) related values to behavior and determine sets of values according to the motivation they foster. Furthermore, Schwartz and Bilsky (1987, 1990) emphasized the fundamental content aspect of a value. What differentiates one value from another is the type of goal or motivational intent. Accordingly, values are not simply abstract conceptions of the desirable, but are motivational (Hitlin & Piliavin, 2004). They represent basic human needs (Rokeach, 1973b; Schwartz, 1992), and these needs drive and motivate social behavior. People can be motivated because they value a particular result in either themselves or others. However, there are three factors worth mentioning which play an important role in discussing the relationship between values and motivation. First, values are certainly not the only motivational factor behind action. Values influence motivation in interaction with other motives, personality traits, attitudes, and situational norms (Staub , 1989). Second, given their abstract nature, values are not related to concrete action like motivational behavior itself (Hitlin & Piliavin, 2004). If a person values performance this does not imply that the same person is acting accordingly and will show a high motivation to perform well. Like the imperfect relationship between attitudes and behavior (Ajzen & Fishbein, 1977), the linkage between values and motivation is thought to be highly mediated. Moreover, motivational behavior may also be influenced by more than one value (Bardi & Schwartz, 2003), which enhances the fact that the relationship between values and concrete motivational behavior is not linear. Third, there is still a need for additional research to clarify whether the affective or cognitive components of values mainly determines the motivational character of a value. According to Bardi and Schwartz (2003), values are supported more by their affective component than by their cognitive component. Maio and Olson (1998) argued that values are socialized into us through the teaching of moral absolutes and are therefore expressions of emotions. However, Williams (1979) supposed "that values are not motivational in an emotional sense, but rather are cognitive structures that provide information that gets coupled with emotion and leads to action" (cited by Hitlin & Piliavin, 2004, p. 380).

Overall, a number of studies suggested that personal values influence motivation in the form of organizational behavior (Ajzen, 1991), goal setting and task performance (Locke & Latham, 1990), and work motivation (Vroom, 1964). Volunteer work, in particular, is one area in which the link between values and motivation has been empirically examined. Benevolence values appear as the most strongly related motivation for volunteer work (Omoto & Snyder, 1995) and are associated with most other measures of volunteer activity. This finding can also be explained by the fact that conviction for a service to the community acts as a value that explains the connection to motivation. Similarly, motivation in the military is mainly driven by a personal conviction regarding the necessity of the military mission and a responsibility for duty and service to the national society. According to Thomas, Dickson, and Bliese (2001), military organizations convey the view that the behavior sets of effective leaders are determined by personal values and motives. Studies have shown that values and motives such as power, affiliation, and motivation to lead are associated with military leader performance (Thomas et al., 2001; Van Iddekinge, Ferris, & Hefner, 2009). This effect was found in addition to personality characteristics such as Extraversion and Conscientiousness. The study by Clemmons and Fields (2011) analyzed the role of values as determinants of the motivation to lead and showed that personal values made significant incremental contributions in explaining all three forms of the motivation to lead (affective-identity, social-normative, noncalculative). Taken together, studies highlighted that as intended by military organizations personal values, along with personality, play a distinct role in predicting motivation.

3.3 Universal virtues: Definitions and theories

As outlined in section 3.1, the integration of virtues into the discourse within philosophy has a long-lasting history. As MacIntyre (1981) confirmed, the concept of virtue has always been an essential explanatory term of scientific interest in moral philosophy.

The term virtue comes from the Latin word *virtus* which is equivalent to "man." The original Greek term is *arete* and means "excellence" (Cawley, Martin, & Johnson, 2000). As mentioned in section 3.1, to the Greeks, a person of *arete* was someone who possessed the life skills to fulfill his or her highest human potential, pointing to the question of *"What kind of person ought I be?"* Given the importance of virtues in moral disciplines, it is not surprising that numerous definitions exist in relation to the concept of virtues. A general definition comes from McCullough and Snyder (2000) interpreting a virtue as "any psychological process that enables a person to think and act so as to benefit him- or herself and society" (p. 1). De Raad and Van Oudenhoven (2011) emphasized that a virtue implies a standard of moral perfection: A virtue is a morally good trait, which serves as a moral beacon and expresses a moral, highly appreciated personality. A more specific definition in the sense of positively valued character traits was provided by Dahlsgaard et al. (2005), defining virtues as "valued human strengths" (p. 203). This definition highlights that virtues are desirable traits, which are worth striving for (De Raad & Van Oudenhoven, 2011). Tjeltveit (2003) points with his definition to the fact that virtues as human qualities are worthy of praise and are relatively stable. Zagzebski (1996) suggested that a virtue as the excellence of the person is directly connected with the idea of excellence and corresponds to a positively evaluated trait of a person. In reference to positive psychology, Yearley (1990, p. 13) defined virtues as dispositions "to act, desire, and feel that involves the exercise of judgment and leads to a recognizable human excellence or instance of human flourishing."

Morales-Vives et al. (2014) sum up the wider definitions by stating that "virtues are desirable personality characteristics that are worth pursuing behaviorally" (p. 298). In alignment with previous studies on interpersonal differences in virtues, in the present thesis a virtue was preferably interpreted as "a moral trait, indicating what one should be or do or show, demonstrate, respect, etc., depending on the form of the term in question" (De Raad & Van Oudenhoven, 2011, p. 45).

3.3.1 Principle studies on identifying and structuring virtues

Combined with the review of numerous publications on virtues over the decades, the analysis of recent research activities confirms the ideas that virtues are a primarily dominant research concept for ethics and philosophy and not for psychology. ISI Web of Science (December 5, 2017) reports 4,117 articles on *virtues* in the field of social sciences. More than half of these manuscripts have reference to philosophy (2,871). The citation report showed that 1,210 publications were assigned to the field of ethics and only 373 articles belonged to psychology.

As discussed in section 2.2, the focus within psychology was dedicated for many years towards the understanding and prevention of mental disorders. As a further consequence, the understanding of moral traits of character has greatly been ignored within the research field of personality psychology (De Raad & Van Oudenhoven, 2011). With the scientific framework of positive psychology, the interest in the study and the classification of virtues has grown because of their importance in different areas of psychology (e.g., Peterson & Seligman, 2004; Sandage & Hill, 2001). To provide an additional argument for the relevance of studying virtues, the link with the work of Gowri (2007) on "corporate virtue" is interesting. He assumed that within the context of internationalization of economies and multicultural collaboration there is an imminent need to discuss consequences for human rights, including virtue issues. Even if business and morality are worlds apart, Gowri (2007) is linking ethics with economic theories and stated that corporate virtues serve the overall aim of sustainable profit and constitute a new model of corporate social responsibility and other-regarding virtues. Within his framework, Gowri (2007) was also highlighting the importance of studies classifying virtues. For example, Fowers and Davidov (2006) discussed the virtue of multiculturalism (that is, the virtue of being open to the other) as an expression of cultural competence.

In addition, Crossan, Mazutis, and Seijts (2013) confirmed that there has been an increasing interest in linking a virtue-related ethical framework with the analysis of different management themes, for example, in reference to executive leadership (Manz, Anand, Joshi, & Manz, 2008), to ethical decision-making in organizations (Arjoon, 2008), to the role of business in society (Arjoon, 2000), and to the educational and pedagogical approach in regards to the doctrine of the economy and business ethics (Mintz, 1996).

Although virtues for many years have not been a primary topic in psychological research, there have been some theories that included this topic in their understanding and empirical research. The following sections refer to three previous studies and more specifically describe their findings on the system of virtues.

a) Eight psychosocial stages by Erikson

Erikson (1963, 1982) was one of the earliest theorists who formulated a classification of virtues in psychology. He adapted Freud's stage theory of development by assuming that the individual person across the life span must confront and accomplish specific social tasks. As a result, different psychosocial virtues develop after each phase. The eight virtues he identified are hope, will, purpose, competence, faithfulness, love, care, and wisdom. One criticism of Erikson's work is the argument that there is little evidence to confirm his severely divided developmental stages and the fact that he believes there is only one proper way to proceed through life (Peterson & Seligman, 2004).

b) Six groups of attributes of moral maturity by Walker and Pitts

Walker and Pitts (1998) examined moral development in Canadian samples. They created a list of 50 prototypical attributes of a very moral person and then asked a group of people to classify these attributes according to similarity. Hierarchical cluster analysis revealed six groups of attributes that describe a high level of moral maturity, namely caring-trustworthy, dependable-loyal, principled-idealistic, confident, being of integrity, and fair.

c) Six core virtues by Dahlsgaard, Peterson, and Seligman

Dahlsgaard et al. (2005) examined human positive characteristics in a historical survey and retraced the theories of philosophers and religious thinkers. As a result, they distinguished six "core" virtues common in these traditions, namely courage, justice, humanity, temperance, wisdom, and transcendence. These core virtues have been accepted not only by moral philosophers and religious thinkers across a broad survey of historical texts and contexts, but they represent almost universally accepted components of what could be widely recognized as representing good character (Peterson & Seligman, 2004).

The attempt of Dahlsgaard et al. (2005) was not meant to be exhaustive and was investigated with explicit reference to the various philosophical traditions. Macdonald, Bore, and Munro (2008) criticized the classification of the six core virtues because they claimed it could not withstand empirical testing. Despite this criticism, Shryack, Steger, Krueger, and Kallie (2010) argued that probably the most systematic approach to the empirical investigation of virtue and character strengths has come from the field of positive psychology, in alignment with the concept of the VIA classification of strengths by Peterson and Seligman (2004). The underlying factor structure of the VIA classification is addressed in the next section.

3.3.2 The factor structure of character strengths and virtues

Peterson and Seligman (2004) did not specify how the theoretical model of the VIA classification should be tested. Since the strengths are being assigned to a virtue, some researchers verified the structure thereof through factor analytic studies (e.g., Peterson, Park, Pole, D'Andrea, & Seligman, 2008; Park, Peterson, & Seligman, 2004a; Peterson & Seligman, 2004; Ruch et al., 2010). However, Peterson and Seligman (2004) pointed out that a person of good character may exhibit one or two strengths within a virtue, but that an individual person seldom shows all strengths within a virtue. Considering the example that an individual does not need to show both high humor and high religiousness in order to be transcendent, the strengths of a virtue do not need to intercorrelate. This illustrates the fact that the anticipated factor structure will not always lead to identify a unique virtue.

In addition, the factor solution loses a relatively large amount of information of the strengths, since factor models are only able to explain a small part (less than half) of the variance of the strengths (Jolliffe & Cadima, 2016).

In other words, factor analyses might not be the suitable method to reliably identify the assignment of strengths to a virtue because factors of strengths do not represent virtues (Ruch & Proyer, 2015). Ruch and Proyer (2015) pointed out that by running factor analysis on the instruments measuring the VIA classification "factors should not be expected to lead to the six virtues in the VIA classification" (p. 11). Furthermore, they stated that a second-order strength factor can be derived from the intercorrelation of the strengths but not a virtue.

Nevertheless, the factorial structure of the VIA-IS generates valuable information regarding a person's character strengths and which character strengths most likely co-occur within individuals. Accordingly, it makes it an important aspect to verify the conclusions and interpretation from data analysis. Specifically, the second-order strength factors identified by data analysis allow to be a more abstract level in assessing an extended view of how strengths (factors) relate to other variables, instead of analyzing the level of single character strengths. In principle, findings which result from factor structure analysis via the VIA-IS varied with respect to the demographical characteristics of samples, with respect to data, being absolute vs. ipsative scores, and are also dependant on the version and language of the VIA-IS. Yet, factor structure research on the VIA-IS delivers three distinct types of findings, differentiating between solutions with two, three, or five factors.

a) Solution with two factors (Peterson, 2006)

Peterson (2006) initiated the discussion about factor analysis based on ipsative[15] data. Two bipolar factors emerged from principal component analysis with oblique rotation, with the strengths being located in a full circumplex. The first bipolar factor was labeled *strengths of the heart* (e.g., kindness, gratitude, humor, and spirituality) vs. *strengths of the mind* (e.g., open-mindedness, persistence, and self-regulation); and the second factor *strengths focusing on the self* (e.g., creativity, curiosity, and vitality) vs. *strengths focusing on others* (e.g., citizenship, fairness, and leadership). The two-factor solution could later be reproduced for the German VIA-IS (Ruch et al., 2010) but not for the Hebrew version (Littman-Ovadia & Lavy, 2012).

b) Solution with three factors (McGrath, 2015; McGrath, Greenberg, & Hall-Simmonds, 2017)

The recent study by McGrath (2015) proposed a three-factor virtue model. Specifically, he identified three virtue factors, labeled *caring, inquisitiveness*, and *self-control*. At the highest level, the structure covered a first component representing good character that split into two components reflecting moral goodness and inquisitiveness (defined as the intellectual interest in the world). The former divided further into components reflecting caring and self-control. The three-factorial structure was confirmed across multiple measures of strengths derived from self-report data corresponding with cultural beliefs about virtue. McGrath (2015) proposed this three-factorial model to be a reliable latent structure for the VIA classification, "encompassing all the elements of positive functioning in society" (p. 418) and capturing "the three targets of virtuous action (others, the self, and the physical world), and so is proposed as an intuitive framework for understanding the conceptualization of character strengths" (p. 418). McGrath et al. (2017) explored this model further and confirmed the substantial

---

15 Ipsative scores are intraindividual standardized scores. They are computed by reducing the individual total mean from each of the scale scores in the VIA-IS.

congruence in three-factor loadings across various samples and the overlap with measures of personality.

c) Solution with five factors (Peterson & Seligman, 2004; Ruch et al., 2010)
Peterson and Seligman (2004) used an exploratory varimax rotated factor analysis of scale scores. The results pointed to a five-factor solution which was similar to the a priori classification of six virtue factors. The orthogonal rotated factors were labeled *interpersonal strengths* (e.g., love, kindness, fairness, and leadership); *emotional strengths* (e.g., bravery, integrity, and vitality); *intellectual strengths* (e.g., creativity, curiosity, and love of learning); *strengths of restraint* (e.g., forgiveness, prudence, and self-regulation); and *theological strengths* (e.g., appreciation of beauty, gratitude, and spirituality). This five-factor structure could also be confirmed for the German version of the VIA-IS (Ruch et al., 2010). McGrath and Walker (2016); Park and Peterson (2006); and Van Eeden, Wissing, Dreyer, Park, and Peterson (2008) applied an oblique factor rotation which led to a four- and one-factor structure, respectively. However, all three studies used the VIA-Youth for exploring the factors.

In a recent study by Ng, Cao, Marsh, Tay, and Seligman (2016) an exploratory structural equation modeling (ESEM) at the item level was applied, using a bifactor analytic approach. The researchers presented the VIA-IS with all 240 character strength items as well as with a reduced set of unidimensional character strength items to a large test sample. It was found that a six-factor structure applied to the reduced set of unidimensional character strength items. The six factors they found were justice, temperance, courage, wisdom, transcendence, and humanity. Additionally, a general factor was defined as dispositional positivity. Given this background, as part of this research, the factor structure of the VIA-IS according to Peterson and Seligman (2004) and Ruch et al. (2010) was adopted for data analysis in the Study II, since the German version of the VIA-IS was used and the five-factor solution could be replicated also in German.

3.4 The psycholexical approach towards the structure of values and virtues

One characteristic of a character strength is its ubiquity, meaning the character strength should be positively valued and receiving recognition across cultures (Peterson & Seligman, 2004). However, different cultures may correspond with different values, virtues, and character strengths. As De Raad and Van Oudenhoven (2008) outlined, different cultures have distinct sets of values. For example, the values of competition and achievement have high importance in countries in which individualism takes a high ranking, whereas cooperation takes high priority in countries with high collectivism (Green & Paez, 2005; Oyserman et al., 2002). Consequently, different structures of values and virtues evolve. That suggested a promising research topic to study the respective structure in different cultures and countries (Morales-Vives et al., 2014). Likewise, the definitions and descriptions of values and virtues vary from one culture to another, since they are part of the cultural context (Sandage & Hill, 2001). Accordingly, values and virtues may not be adequately assessed by applying the same measuring instrument in different cultures.

The psycholexical approach offers a very useful method to reveal these cultural differences. According to Goldberg (1981), language will represent the most significant individual differences in daily transactions with other people. The lexical hypothesis by Goldberg (1981) assumed that people wish to talk about what is important to them and that the words they use for this purpose are found in the lexicon. In line with this understanding, the so-called psycholexical approach takes the lexicon in its particular language as the starting point to identify the descriptors of human behaviors. As such, the lexical approach does not depend on the accuracy of preexisting scientific concepts (De Raad, 2000). The psycholexical approach enables the researcher to identify a comprehensive set of descriptors of the psychological concept of interest. Originally, it has its roots in personality psychology to develop the Big Five-factor model and has traditionally been used to classify personality traits (cf. De Raad & Mlačić, 2015). Angleitner, Ostendorf, and John (1990) applied the psycholexical approach to various person–characteristics, including a value-related class of terms called "attitudes and worldview" (cf. Saucier, 2000). Other application of the psycholexical program of research was done on interpersonal behavior (De Raad, 1995). Psycholexical studies on values and virtues followed. This area of research initiated a new method of empirical examination on values and virtues, and selected studies will be addressed in more detail in the following section.

3.4.1 Psycholexical studies on universal values

Aavik and Allik (2002) introduced the iBa to apply the psycholexical principles to the domain of values. The aim was to develop an exhaustive and culturally sensitive list of universal values in the Estonian language. A series of researchers followed this novel study design immediately, e.g., Renner (2003b; for the German language in Austria); Renner, Peltzer, and Phaswana (2003; for the Northern Sotho language in South Africa); Renner and Myambo (2007; for the Egyptian language); and De Raad and Van Oudenhoven (2008; for the Dutch language in the Netherlands). In subsequent studies Morales-Vives et al. (2012;

for the Spanish language) and Crețu et al. (2012; for the Romanian language) conducted further psycholexical studies on values.

In review, the study by Aavik and Allik (2002) selected a set of 560 words from the Estonian lexicon and in subsequent steps drastically reduced them to a final list of 78 value descriptors. These expressions were administered to 294 participants with the instruction to indicate the extent to which the values were guiding principles in their lives. Principal component analysis revealed six factors, labeled *Self-enhancement, Self-realization, Hedonism, Benevolence, Broadmindedness,* and *Conservatism*. The analysis carried out in this study revealed that the constructs measured by the SVS and the value factors are only partially interchangeable. Yet, the general structure of the extracted expressions corresponded with the two-dimensional level of higher-order values described by Schwartz (1992). Therefore, Aavik and Allik (2002) reasoned that in addition to an universal set of core values measured by the SVS, there is a more specific set of values for a given culture to be captured by a psycholexical analysis.

Within Austria, Renner (2003b) made the distinction between instrumental values and terminal values. He identified 383 nouns and 299 adjectives from the German lexicon and administered the two lists to 456 participants. Principal component analyses were performed, yielding four instrumental value factors and five terminal value factors. The four instrumental value factors were basically the same as four of the five terminal value factors. For this reason, Renner made the conclusion that individuals do not distinguish between instrumental and terminal values. The maximum set of the five factors were labeled *Profit, Balance, Salvation, Intellectualism,* and *Conservatism* (which was the additional terminal value factor). Further, the new instrument called Austrian Value Questionnaire (AVQ; Renner et al., 2004) was developed on this lexical basis and is used as a culture-specific measure of value orientation for German-speaking countries.

Renner, Peltzer, and Phaswana (2003) aimed at establishing a taxonomy of values in Northern Sotho within South Africa. The extracted list contained totally 210 value-describing terms. A total of 400 participants rated these expressions with regard to their subjective importance as guiding motives in life. Principal component analysis yielded five factors, defined as *Religiosity and Support, Solidarity, Conformity and Benevolence, Leadership and Achievement,* and *Human Enhancement*. Cross-cultural comparisons with the psycholexcially derived list in German (Renner, 2003b) pointed out that in Northern Sotho, religious themes and social commitment play a more important role than in German, and mainly reflect the religious and collective values of the traditional African culture.

De Raad and Van Oudenhoven (2008) carried out the study in Dutch and first identified 4659 value-relevant expressions. The list was subsequently reduced until it contained 641 values which were administered to 634 participants to rate the extent to which they were guided by the values in their life. A principal component analysis of the list of 641 lexical words yielded eight factors. The identified factors are *Status and Comfort, Organization and Achievement, Competence, Love and Happiness, Benevolence, Spirituality, Family and Tradition,* and *Aesthetics and Erudition.*

Morales-Vives et al. (2012) determined the structure of values in the Spanish population on the basis of the psycholexical approach. They initially identified 2356 value-describing words and reduced them to a final list of 566 words. Subsequently, they had these expressions rated by 532 Spanish university students relating the extent to which they were guided by each value. Principal component analysis revealed seven factors: *Social Recognition, Competence, Love and Happiness, Benevolence, Idealism, Equilibrium,* and *Family.*

3.4.2 Psycholexical studies on universal virtues

In the domain of virtues, Cawley and colleagues (2000) initiated the application of the psycholexical approach to explicitly study virtues, followed by the Dutch researchers De Raad and Van Oudenhoven (2011) and the Spanish study by Morales-Vives et al. (2014). Specifically, Cawley and colleagues (2000) developed the Virtues Scale (VS), utilizing the heuristic criteria "What ought I to be?" and "What ought I to do?" to determine whether the descriptive terms within the dictionary were suitable for describing virtues. Combining adjectives and nouns resulted in a list of 140 extracted virtue terms. Subsequently, the virtues were converted into the VS items, representing indicators of moral personality. Each virtue was described by means of three different statements to further specify the meaning of the abstract virtue descriptors. The items were administered to American participants with the instruction to indicate the extent to which the items describe "how they really are, and not how they ideally should be." Based on the ratings a factor analysis with varimax rotation was performed and resulted in four factors. They were labeled *Empathy, Order, Resourcefulness,* and *Serenity.*

The psycholexical study by De Raad and Van Oudenhoven (2011) in Dutch included the exhaustive list of 1,203 personality traits developed by Brokken (1978). Eleven judges reliably selected 153 virtue descriptors from this list. Subsequently, a total of 400 participants (200 pairs) provided ratings on these 153 terms: One member of a pair provided a self-rating, and the other rated his or her partner. Principal component analysis yielded six factors: *Sociability,*

Achievement, *Respectfulness*, *Vigor*, *Altruism*, and *Prudence*. This was the first study to combine self- and peer-ratings in the psycholexical study of virtues and values. The results indicated that peer-ratings were less friendly and helpful (Sociability), less well-mannered and polite (Respectfulness), and less self-assured, courageous, and optimistic (Vigor) than the perception by other people. However, the factor structure of the self- and partner-ratings were similar as measured through congruence coefficients.

Recently, Morales-Vives et al. (2014) identified 209 virtue-describing terms on a psycholexical basis. The descriptors were administered to 485 participants with the instruction to rate the extent to which each virtue term applied to them. By means of a principal component analysis, they determined a dimensional structure of seven factors: *Self-confidence*, *Reflection*, *Serenity*, *Rectitude*, *Perseverance and Effort*, *Compassion*, and *Sociability*. The authors used self-ratings only and concluded that the seven Spanish factors are comparable to the factors identified in other cultures, although there is no full equivalence when comparing the labels, definition, and content of the factors.

3.5 Classifications of values and virtues: Summarizing overview

This section reviews the existing systems of values and virtues, as reported through research articles, specifically the ones referenced by De Raad and Van Oudenhoven (2008) and Morales-Vives et al. (2012) on values, as well as De Raad and Van Oudenhoven (2011) and Morales-Vives et al. (2014) on virtues.

Table 4 provides an overview of the different systems of values, whereas Tab. 5 compiles the existing systems of virtues. Two principle remarks apply: First, cross-cultural comparisons of structures in personality characteristics show both commonalities and cultural-specific results (e.g., Saucier, Hampson, & Goldberg, 2000). As mentioned above, various studies have been conducted to compare important factors of values and virtues between different cultures (e.g., Wiggins, 1991; Hofstede, 2001; Schwartz, 1999). The concepts of agency-communion (Wiggins, 1991) and individualism-collectivism (Hofstede & Bond, 1984) represent common two-dimensional concepts to capture individual differences in values and virtues. These concepts are also known as "deep structure" of cultural differences (Greenfield, 2000). They have been statistically confirmed in most of the mentioned studies. For instance, the factors *Sociability*, *Prudence*, and *Respectfulness* were typical of a communion-orientation, while *Achievement* and *Vigor* were found to be of an agency-orientation, as shown within the virtue structure of De Raad and Van Oudenhoven (2011). Second, the different value and virtue systems differ in terms of number and descriptive naming of factors,

Tab. 4: Overview of the systems of values (adapted to the version of Morales-Vives et al., 2012)

Spranger (1928) **6 factors**	Schwartz (1992) **10 factors**	Aavik & Allik (2002), psycholexical, Estland. **6 factors**	Renner (2003b), psycholexical, Austria. **5 factors**	Renner, Peltzer, & Phaswana (2003), psycholexical, South Africa. **5 factors**	De Raad & Van Oudenhoven (2008), psycholexical, the Netherlands. **8 factors**	Morales-Vives, de Raad, & Vigil-Colet (2012), psycholexical, Spain. **7 factors**
Power type vitality; perseverance; self-realization	*Power* wealth; authority; preserving; public, image	*Self-enhancement* power; ambition; self-promotion	*Profit* wealth; possession; career	*Leadership and achievement* wealth; competition; perseverance; leadership; pride;	*Status and comfort* wealth; property; beauty; reputation; success; reward	*Social recognition* popularity; leadership; power; prestige; social status
Economic type work; production; security	*Achievement* intelligent; ambitious; successful	*Self-realization* independence; wisdom; experience			*Organization and achievement* conscientiousness; structure; discipline; order; punctuality; industriousness; concentration	
	Self-direction choosing own goals; creativity; independent			*Human enhancement* self-control; free will; responsibility; joy; pleasure; humanity; piety	*Competence* decisiveness; self-assuredness; enterprising spirit; change; autonomy	*Competence* intelligence; wisdom; education,

(continued on next page)

Tab. 4: Continued

	Hedonism	Hedonism	Stimulation	Benevolence	Benevolence	Balance	Solidarity	Love and happiness	Love and happiness
	pleasure; enjoying life	excitement, sexuality; consumption					cooperation; friendship; alliance; relationship; attachment	spontaneity; cheerfulness; being sensitive; interpersonal warmth; tenderness; enthusiasm; friendliness	being friendly, cheerfulness, holding others in esteem
Social type empathy, loyalty; love			varied life; exciting life; daring	forgiving, helpful; true; friendship	helpfulness; sincerity; kindness	fairness; trust; human rights	Conformity and benevolence honor; care; protection; cooperation; forgiveness	Benevolence being accommodating; big-heartedness; being mild-mannered; charity; peacefulness	Benevolence solidarity, hospitality

Religious type salvation; relativity of human existence	Universalism world at peace; broadminded; unity with nature	Broadmindedness tolerance; humanity; inner peace	Salvation faith in god; piety; religion	Religiosity and support Christianity; purity; care/caution; saving; salvation; strength	Spirituality religion; praying; god; morality; nature	Equilibrium harmony, comfort, relaxation
Theoretical type rules; principles	Conformity honoring parents; politeness; self-discipline	Conservatism order; neatness; decency	Conservatism national identity; tradition; duty		Family and tradition marital duty; marriage; parental duty/love; family ties; parenthood	Family family life, family ties; job security
	Tradition detachment; moderate; respect for tradition					
Esthetic type individual expression; beauty	Security clean; family security; social order		Intellectualism reflection; open-mindedness; culture		Aesthetics and erudition art; profundity; artistry; art of living; originality; creativity; culture	Idealism Ideology; nationalism; patriotism

even though the factors refer obviously to the same scope (cf. De Raad, Morales-Vives, Barelds, Van Oudenhoven, Renner, Timmerman, 2016).

Referring to universal values, Tab. 4 summarizes the structures by Spranger (1928), Schwartz (1992), and the five psycholexically derived structures (Aavik & Allik, 2002; De Raad & Van Oudenhoven, 2008; Morales-Vives et al., 2012; Renner, 2003b; Renner, Peltzer, & Phaswana, 2003) making a distinction between five to eight universal value factors. Strong evidence suggested the existence of some common values in different systems having the same label. The most obvious example thereof was *Benevolence*, defined by Schwartz (1992), Aavik and Allik (2002), De Raad and Van Oudenhoven (2008), and Morales-Vives et al. (2012). Some value factors were consistent in content and possess a different label. For instance, the factor labeled *Power* by Schwartz (1992) was called *Status and Comfort* by De Raad and Van Oudenhoven (2008), and *Self-enhancement* by Aavik and Allik (2002). Other values seemed specific to the culture and accordingly could not be easily replicated (e.g., *Aesthetics and Erudition* from the De Raad and Van Oudenhoven study).

Table 5 depicts the structures by Erikson (1963, 1982), Walker and Pitts (1998), Dahlsgaard et al. (2005), and the three psycholexically derived structures (Cawley et al., 2000; De Raad & Van Oudenhoven, 2011; Morales-Vives et al., 2014). All the systems contained four to eight factors of universal virtues. In fact, they showed similarities in content. For example, the virtue factor *Wisdom* from the systems of Erikson (1963, 1982) and Dahlsgaard et al. (2005) was similar to the Dutch factor *Prudence* (De Raad & Van Oudenhoven, 2011) and the Spanish factor *Serenity* (Morales-Vives et al., 2014). However, they did not have a clear equivalence in the content of the other virtue systems. The virtue factor *Love and Care* by Erikson (1963, 1982) seemed to be represented by equivalent factors in all the other systems, although they were variously labeled and range from *Caring-trustworthy* (Walker & Pitts, 1998) to *Compassion* (Morales-Vives et al., 2014). Likewise, the Dutch factor *Achievement* had an equivalent factor in most of the other systems. However, *Achievement* was not covered in the Dahlsgaard and colleagues (2005) system, nor is the Dutch factor *Respectfulness*. In general, the Dutch and the Spanish virtue systems were the ones that have widest representation in the other virtue systems. In fact, most of the factors from the other systems were covered in the Dutch and Spanish systems to different degree.

In summary, the interest within the scope of this research is in the psycholexical approach, which has proven to be a useful methodology to identify cultural differences in values and virtues. The psycholexical studies performed in different countries revealed between five and eight value factors, and four and seven virtue factors. The findings illustrated the assumption by De Raad and

Tab. 5: *Overview of the systems of virtues (adapted to the version of Morales-Vives et al., 2014)*

Erikson (1963, 1982) 8 factors	Walker & Pitts (1998) 6 factors	Dahlsgaard, Peterson, & Seligman (2005) 6 factors	Cawley, Martin, & Johnson (2000), psycholexical, USA. 4 factors	De Raad & Van Oudenhoven (2011), psycholexical, the Netherlands. 6 factors	Morales-Vives, de Raad, & Vigil-Colet (2014), psycholexical, Spain. 7 factors
Hope trust, gratitude		*Transcendence* gratitude; hope; spirituality	*Serenity* serene; peaceful; merciful	*Sociability* friendly; social; good-hearted; kind	*Sociability* love, happiness, tenderness, communicative, comradeship
Love intimacy	*Caring-trustworthy* honest; truthful; good; caring; kind; helpful	*Temperance* forgiveness; humility; prudence	*Empathy* sympathy; understanding; compassion	*Altruism* sacrificial; compassionate; noble	*Compassion* humanitarian; generosity; mercy; generosity
Care kindness; generativity	*Dependable-loyal* responsible; honorable; faithful; loyal; respectful	*Humanity* love; kindness	*Order* discipline; scrupulous; tidy	*Respectfulness* orderly; tidy; obedient; civilized; decent	*Rectitude* seriousness; decency; discipline; integrity; politeness
Fidelity identity; loyalty	*Principled-idealistic* law-abiding; clear values; principled; self-disciplined	*Courage* bravery; persistence; integrity	*Resourcefulness* perseverance; purposeful; fortitude	*Vigor* decisive; vigorous; brave; will-power	*Self-Confidence* drive; vigor; bravery; strength
Competence industry	*Confident* strong; self-assured	*Wisdom* creativity, curiosity; open-mindedness; perspective		*Achievement* tenacious; industrious; diligent; dutiful	*Perseverance and Effort* responsibility; hard-working; effort; commitment
Will autonomy; determination	*Integrity* consistent; hard-working; conscientious	*Justice* fairness; leadership; citizenship		*Prudence* integrity; discrete; philosophical; open-minded; sober-minded	*Serenity* calm; tranquility; peaceful; patience; caution

(continued on next page)

Tab. 5: Continued

Purpose	*Fair*
initiative,	virtuous; fair;
courage	just
Wisdom	
integrity,	
perspective	

Van Oudenhoven (2011), that different cultures correspond with distinct sets of values and virtues. However, due to the language- and culture-based differences within the descriptive psycholexical approach, there are limitations to comparing the findings in reference to the factor structure of values and virtues. This corresponds with Berry (1969) who pointed out that it is not possible to transfer the results from one language to the other. To allow cross-cultural comparability, De Raad and Van Oudenhoven (2011) stated "It is crucial that similar psycholexical procedures are applied in a variety of other languages/cultures" (p. 45).

It needs to be mentioned that in previous studies the psycholexical approach was primarily applied to explore the variation of value and virtue structure in different language-related cultures. However, this thesis strongly supported the assumption that the psycholexical approach is equally qualified to assess values and virtues under the condition of different organizational cultures. This is in line with De Raad, Timmerman, Morales-Vives, Renner, Barelds, and Van Oudenhoven (2017) pointing out that the psycholexical approach is a useful method in exploring the field of values and virtues in different cultures. This confirms the principle conclusion that the psycholexical approach can be adapted to any organization with its distinct culture, including the military. Accordingly, this thesis aligned the psycholexical approach with the specific culture of the military environment and followed the methodological principles as illustrated in the Dutch studies of De Raad and Van Oudenhoven on values (2008) and virtues (2011). The ultimate aim was to develop a set of value and virtue descriptors within the Swiss Armed Forces and to assess the structure of Swiss military values and virtues.

3.6 Conceptual difference between values and virtues

In ancient times, a distinction was already being made between values and virtues, as mentioned above in section 3.1. Similarly, the subject of values and virtues is treated conceptually separately in the latest studies (e.g., Morales -Vives et al.,

2014). At the same time, values and virtues have in common that "they refer to relatively stable characteristics of individuals" as pointed out by De Raad and Van Oudenhoven (2011, p. 44). Both values and virtues shape the way people behave in their daily lives. Furthermore, there are clear differences between these two concepts, with values defined as *desirable goals* that serve as guiding principles for individuals (Schwartz, 1992) and virtues as *desirable* character *traits* which are positively valued (De Raad & Van Oudenhoven, 2011; Dahlsgaard et al., 2005). The addition of *valued* in the definition of virtues illustrates the conceptual tie between values and virtues in the sense of something valuable and worth striving for. Consequently, values and virtues can be seen in a broader context as related to each other, even though they are not specifically related to the same thing.

Likewise, the separation by Rokeach (1974) into terminal and instrumental values implied the conceptual difference between values and virtues: Terminal values are abstract nouns and can be considered as value-related, whereas instrumental values correlate well with concrete modes of conduct as norms of behavior and show a link to virtues.

Furthermore, virtues are a subset of traits that *dispose* people to behavior and values *convey* what people find important in a sense of moral preference (cf. Morales-Vives et al., 2014). Accordingly, values provide direction to behavior as a kind of moral compass, but they are not regarded to dispose people to behave in a certain way as traits do. Additionally, values can be prioritized in a hierarchical preference given to certain values over others. Consequently, people can have particular values but their actual behavior may not be aligned with their values. In addition, they could value traits they do not have or not value traits they do have. For instance, if a person is characterized as loyal to his supervisor, this means that this person is expected to behave in a loyal way. If a person has discipline as a value, this does not necessarily mean that he is a disciplined person; he can only expect it from his employees and possibly will select different courses of action in his role as a supervisor.

Occasionally, it can be rather challenging to distinguish between values and virtues. There are lexical terms which overlap in practical usage and which may fulfill both functions of a value and a virtue (e.g., honesty, kindness). However, most terms are either a value or a virtue; at the same time, both are perceived as ethical terms. Values and virtues both have a positive connotation and are regarded as desirable (Van Oudenhoven, De Raad, Carmona, Helbig, & Van der Linden, 2012). Consider, for instance, terms such as *affected* or *distrustful*, which have negative connotations, and do not describe either a value or a virtue. These terms are dispositional traits, but do not involve excellence in their meaning nor are they regarded as desirable (Hampson, Goldberg, & John, 1987).

Crossan et al. (2013) explicitly pointed to the connection of values with virtues by supposing that virtues unlike other personality traits are characterized by their intrinsic value. Peterson and Seligman (2004) brought it to the point: "Virtues embody values when the behavior they organize and direct becomes habitual" (p. 74). They provided an overview of the linkage of the ten universal values by Schwartz (1992) with their corresponding character strengths in the VIA classification (e.g., Achievement as a value with persistence as a strength; Power as a value with leadership as a strength). However, they state that there is only a very inaccurate correspondence between a universal value and a character strength. Boe (2017) refers to a similar understanding as Peterson and Seligman (2004) and assumes that "a person can express his or her values through one's character" (p. 114).

In a nutshell, virtues and values are two concepts that are not the same, yet are sometimes treated as if they were. Olsthoorn (2011) observed that in the practical application of these concepts in the military environment, it is especially noticeable that values and virtues are perceived as the same. In the theoretical understanding of military ethics, Robinson (2008) stated that virtues represent "desirable characteristics of individuals, such as courage," while values correspond to "the ideals that the community cherishes, such as freedom" (p. 5).

4 Research on military values and virtues

As described in section 2.3, there is an increasing number of studies in the field of positive psychology showing that values, virtues, and character strengths are important in the military context of work satisfaction, individual adaptation ability, performance, and effective leadership. In addition, there is growing evidence that understanding soldier performance requires addressing both cognitive and non-cognitive factors (Böhm, 2008). More specifically, Matthews, Lerner, and Annen (2019) were referring to the finding by Neisser et al. (1996) and stated that – beyond measures of talent such as intelligence, aptitude, and other cognitive constructs – measures of non-cognitive factors, i.e., values, virtues, character strengths, and personality, appear to account significantly in variation of human performance. This specifically applies when considering the military environment, where decisions are often made under extremely challenging conditions, including sleep deprivation, and fear of death or severe bodily harm (Kornguth et al., 2010).

As Matthews (2014) stated, "asymmetric" warfare is a specific challenge to modern day soldiers. This type of war, characterized by terrorism, guerrilla warfare, and ideological manipulation, constantly forces the soldier to rethink the role, norms, and values in the military profession (Snider & Matthews, 2012). Linked with this thought was the question of what are the typical values and virtues within a military organization and which factorial structure can be identified. In the following section, important findings on the factor structure of values and virtues within the military context are addressed.

4.1 Significant findings on the structure of military values and virtues

Military organizations differ from civilian institutions with respect to the specific culture, being identified by a professional commitment of military persons working for the greater good of the society (Hillen, 1999; Proyer et al., 2012). Bachman, Sigelman, and Diamond (1987) found that people who intend to join the military differ in their values and attitudes from those people who have reasons not to join. Franke (2001) found that persons considering a career in the US military tended to be more conservative and patriotic. Additionally, there is evidence that military experience affects the life course (Settersten, 2006), personality traits (Jackson, Thoemmes, Jonkmann, Ludtke, & Trautwein, 2012), political attitudes, and values (Jennings & Markus, 1977). However, the effects are statistically small when considering relevant control variables, e.g., for attitudes prior to entering military service.

A further relevant study by Matthews et al. (2006b) demonstrated the distinct nature of military culture and its influence on the character of the military persons. They compared character strengths of three samples (West Point, Norwegian Naval Academy cadets, and US civilians) and showed that the rank order of the character strengths of the American West Point cadets was more consistent with the military sample in Norway than with the civilians in their own country.

Thus, taking the assumption that military culture differs from civilian culture, very little empirical research has focused on what personality characteristics are associated with military service. Matthews (2014) stated that successful military personnel must be "leaders of character" (p. 370), and they need to be transparent, fair, and honest in interacting with others. They also need to be morally courageous and loyal to their country and comrades.

The following studies help identify the ways in which military culture shapes the values and virtues of military personnel, and illuminates the underlying structure of the military values and virtues.

4.1.1 Findings on military values

Recent research on military values focuses on the fundamental question of how to ensure a desired military culture by means of selection and socialization (for an overview, see Jackson et al., 2012). Corresponding studies examined the topic of values or value change in the US military (e.g., Bachman et al., 1987), as well as in Germany (Jackson et al., 2012), South Africa (Franke & Heinecken, 2001), Australia (McAllister & Smith, 1989; McAllister, 1995), and Japan (Inagaki, 1975). Thus, there has been little research about the structure of military values which could help to specifically describe the value culture of a military organization and to better understand the influence of cultural components on character development of recruits and on external variables such as leadership performance and job satisfaction.

Schumm et al. (2003) investigated the structure of military values in the US Army. They used various measures of professional military values to assess structure and how it relates to military outcome variables such as retention, morale, satisfaction with the quality of Army life, overall job satisfaction, and wartime preparedness. They administered a questionnaire of 15 value-related items to 7,860 soldiers in the US Army. The result of the factor analysis based on the collected ratings was a four-factor solution, i.e., (1) Military Dedication (loyalty to the US Army, to the military unit, the nation, and to a willingness to risk one's life for the military organization); (2) Integrity (being honest and doing what is right); (3) Military Bearing (maintaining a military appearance, high moral standards on- and off-duty, and showing proper respect and military courtesy to others); and (4) Job Commitment (job dedication, personal drive to succeed, and being committed to working as a team). Military Dedication exhibited the largest correlation with outcome variables such as individual morale, retention intentions, and deployment readiness. Furthermore, high correlation was found between military values and selected demographic variables (i.e., rank, years of service, age). The authors therefore suggested the inclusion of multivariate analysis based on value factors, demographic variables, and outcome variables. The four scales of military values derived in this study showed very high internal consistency as well as significant correlations with important outcome variables with particular interest to military researchers. However, this study was not based on a psycholexical approach which would have required that a list of military relevant values was derived and agreed upon across the organizational hierarchy.

Another study conducting structural analysis is the one by Britt, Stetz, and Bliese, (2004) using the RVS by Rokeach (1973b) to assess universal values among US Army rangers, an elite unit within the Army component of the US Armed Forces. They applied a principal axis analysis on the basis of the 18 instrumental values to explore their structure. The analysis supported to retain two factors, one accounting for 49% of the variance in items and one accounting for 8%. The two distinct factors that emerged were named Achievement Values and Affiliation Values. Examples of values loading on the Achievement factor included ambitious, capable, courageous, honest, imaginative, and independent. Examples of values loading on the Affiliation factor were forgiving, loving, cheerful, and polite. It needs to be mentioned that this study was conducted with an instrument to measure universal values but not with a measure which is particularly conditioned to the military, such as a psycholexical derived questionnaire to assess the cultural-specific aspects of this specific sample.

Also, a further remark relates to the fact that the study by Schumm et al. (2003) and the one by Britt et al. (2004) was conducted more than 10 years ago. Any more recent literature pointing to similar studies about the empirical factorial assessment of military values was not found.

In the field of describing differences in value orientations among national groups of officers, however, further studies on military values are existing and worth mentioning. For instance, the research of Franke and Heinecken (2001) used the value factors of Conservatism, Patriotism, Warriorism, Globalism, Peacekeeping, and Individualism vs. Collectivism, showing that officers of the South African Military Academy and the US Military Academy at West Point differed significantly in their value factors. Soeters (1997) presents the results of a comparative study in military academies in thirteen nations. The study used the four Hofstede dimensions (power distance, uncertainty avoidance, individualism, and toughness; see section 3.2.1) to measure values and the perception of individual members of the organization, computing an aggregated score on organizational and national level. The results indicated that there are large differences between military cultures in different nations. Interestingly, the distribution among the various military academies on all cultural factors was even larger than in the original Hofstede study with parallel civilian samples. For instance, the difference between the Canadian academy and the Italian academy is quite systematic. The military cultures in these two academies represent extreme types as the beginning and the end of a process of development. The author assumed that the Italian and Belgian academies are in the initial stage, i.e., at the traditional side of the continuum with a cultural profile "which is relatively (machine) bureaucratic and institutional" (p.179) and a high degree of power distance, whereas the academies in Canada, USA, Denmark, and Norway represent the more contemporary or even the future form of military life with a profile characterized as "professional bureaucratic and occupational" (p.179) and a low degree of power distance.

In addition to the culture-related items, the study of Soeters (1997) included a number of questions with respect to the importance of military discipline. For the Danish and Canadian officers some aspects of formal discipline, such as salutes or ceremonies, are considered important, while they are valued as less important in Germany, the Netherlands, Belgium, and Norway[16]. This finding is remarkable because Denmark and Canada are countries that represent a professional military culture, as described above. Consequently, the author assumed that a contemporary military culture such as a professional culture does not deny the importance of formal discipline. On the basis of his analysis, Soeters (1997) concluded that military academies and armed forces with modern instead of traditional cultures would be better positioned to cope with current circumstances and requirements.

Battistelli, Ammendola, and Galantino (1999) referred to the subject of changing values in course of a military career. They assumed that with the increasing age of soldiers there is a tendency to recognize a shift towards traditional values such as Security. The authors presented the possible hypothesis effect related to the socialization of the members of the organization: Senior officers have undergone a learning process that leads to full alignment with basic cultural beliefs, and to internalizing an attitude of consciously sharing organizational values.

4.1.2 Findings on military virtues

Virtues have always been a prominent topic in the context of military ethics (Aronovitch, 20001; Davenport, 1986). Much has been written on virtue ethics from a philosophical perspective. This tradition goes back at least as far as Aristotle. The central military virtue seems to be courage as described in the Nicomachean Ethics (Aristotle, trans. 2000, 1115a): "In the primary sense, then, the courageous person will be said to be the one who is fearless about a fine death, or about sudden situations that threaten death; and those that occur in war is mostly of this sort." A later virtue ethicist thinker Samuel Huntington mentioned loyalty and obedience as "the highest military virtues" (1957, p. 73).

Olsthoorn (2013) wondered if the changes in the military's wider environment with the shift from traditional tasks to new and more complex missions would lead to new questions, such as whether some military virtues have lost their importance, and have others perhaps gained significance?

---

16 This finding is referring to the time before 1997, when Germany, the Netherlands, Belgium, and Norway were organized in a system of conscripts.

As Olsthoorn (2013) outlined, there is a new attention directed towards virtue ethics and military virtues in the education for military personnel. The main reason is attributed to the focus on character building for military personnel rather than imposing rules of conduct from superior leadership levels (Olsthoorn, 2011). There is a growing willingness within military organizations to consider the idea that character and therefore virtue can to some extent be shaped and strengthened. However, the existing literature is surprisingly scarce. Most studies deal with one specific virtue only, such as courage or loyalty, but extensions into wider relations between different virtues are rare.

Although Peterson and Seligman's character strengths and virtues fit well with the military's emphasis on character, research focusing on the profile of character strengths and virtues among military samples is still in its infancy. Matthews et al. (2006b) conducted what is believed to be the first empirical assessment of the 24 character strengths and virtues in a military context. They administered the VIA-IS to the three different groups, consisting of 103 West Point cadets, 141 Norwegian Naval Academy cadets, and 838 comparable US civilians (i.e., at the age of 18–21, with some college). The highest character strengths among the two military samples were honesty (integrity), hope, bravery, industry (persistence), and teamwork. The majority of these top character strengths (honesty, bravery, and industry) lay within the virtue domain of courage, consolidating emotional strengths that involve the exercise of will to accomplish goals in the face of opposition. Matthews et al. (2006b) pointed out that this is consistent with values described in the Army doctrine. Teamwork falls under the domain of justice, which encompasses civic strengths that underlie healthy community life, while hope is within the domain of transcendence, entailing strengths that forge connections to the larger universe and provide meaning (Seligman, Steen, Park, & Peterson, 2005). It is worth noting that leadership was not even in the top half of the ordered character strengths in the military samples. It accounts for 15 out of 24 (along with spirituality) for the West Point cadets and tied for 17th place (along with forgiveness) for Norwegian Naval Academy cadets. However, it is unsurprising that 3 of the top 5 (signature) strengths were in the virtue domain of courage. A rather surprising finding in Matthews et al.'s study was that kindness (7th), humor (8th), love (11th), and gratitude (12th) were all higher than the character strength of leadership in the West Point cadet sample. Furthermore, citizenship, persistence, integrity, and bravery seemed to be important for military success in both samples of West Points and Norwegian cadets.

At West Point, a five-year longitudinal study of character assessment and development, called *Project Arete*, has been recently launched (Callina et al., 2017). Conducted by scientists from Tufts University, West Point, and a variety of other institutions, *Project Arete* will allow researchers to systematically explore the measurement and development of character in a sophisticated, longitudinal design. Similar studies conducted in other institutions and contexts are needed to better understand how to measure character and how to design educational systems to enhance it.

Matthews and colleagues (2006b) stated that it is important to extend the criterion-based validity of character strengths and virtues research to other operationally significant military settings. For instance, Gayton and Kehoe (2015) studied Australian Army Special Forces applicants, asking them to rank themselves on 24 character strengths at the beginning of the selection process. Across all applicants, the character strength of integrity was most frequently assigned a top-four rank (45%), followed by teamwork (41%), persistence (36%), and love of learning (25%). Successful applicants were assigned a top-four rank to teamwork significantly more often than unsuccessful applicants (65% versus 32%).

Cosentino and Solano (2012) examined the differences in positive traits between Argentinean military and civilian college students and between cadets in their first and final years at military academy. Additionally, they studied the relations between positive traits and the academic and military performance of Argentinean cadets in their first and final years, according to the classification of positive traits by Peterson and Seligman (2004). The results generally showed that when age and career stage were held constant, the scores of the military students were higher than the scores of the civilian students across various strengths. Military students reported higher levels of the character strength of spirituality than did civilian students. The relationships between strengths and performance differed for students in their first and final years at the military academy. In particular, cadets with the higher levels of academic or military performance in their last year (i.e., the cadets best adapted to the academy) reported higher levels of the character strength of persistence when compared to low-performing cadets in the same year of study. Eggimann and Schneider (2008) studied character strengths and virtues of Swiss career officers and found bravery, persistence, vitality, citizenship, leadership, self-regulation, hope, and curiosity as signature strengths of this specific professional group (compared with the norm sample of Swiss civilians).

A virtue or a value is usually a positive personal characteristic of an individual person. Accordingly, this thesis was conceptualizing values and virtues as individual characteristics of personality (see section 3.6). However, one can also have a perspective on values and virtues as a collective concept. Sandin (2007) believed the military to be suitable for ascriptions of collective virtues: "The military contains good examples of groups to which collective virtue can be reasonably ascribed" (p. 308). This understanding can also be adjusted to values. Several authors emphasized that military activities and the ethics involved in them are collective enterprises. For example, Huntington declared: "The officer submerges his personal interests and desires to what is necessary for the good of the service [...] The military ethic is basically corporative in spirit. It is fundamentally anti-individualistic" (1957; p. 63).

As Matthews (2012) stated, it is necessary for a military organization to understand its own culture and the underlying core values and virtues. In accordance, Britt et al. (2006) pointed to the importance to define a classification of values and virtues within the military organization. It is therefore of interest how military organizations effectively handle this requirement.

4.2 International practical perspective

Given the current research finding on military values and virtues, it was of broader interest to learn about other military organizations regarding their principles and practical way of dealing with military core values and virtues. The approach was with the intent to extend the view beyond the Swiss Armed Forces. A questionnaire was established to ask organizations in other countries if a classification of values and virtues existed within their respective organizations. The typical questions were as follows:

- Is there a list of core values and/or virtues in practical use as a binding behavioral directive in your organization?
- How many values and/or virtues does your list include?
- Please fill in the corresponding values and/or virtues that are part of this list.
- What is the official name of this list?
- What is the overall objective in making use of the list in your organizational unit? Participants could choose one or several of the following points: Leadership and management, training and education, strategy-setting, and communication.
- Based on your professional experience, do you consider that the classification of values and/or virtues has a positive impact on leadership, training and education, strategy-setting, and communication?

To capture the anticipated data, 19 nations were included in a descriptive online study, using the network of the International Military Testing Association (IMTA). The participants were instructed to provide their answers with reference to the specific organizational division in which they are working. Furthermore, they were informed about the conceptual differences between the values and virtues and advised to answer the questions by using the same wording terminology as practiced in their organization.

Feedback from 15 nations was received, corresponding to 39 representatives working in military organizations, occasionally within the same country but in different organizational units. The responding nations were: Australia, Austria, Belgium, Canada, Estonia, France, Germany, India, Indonesia, the Netherlands, Norway, Singapore, South Africa, Sweden, and United States. The following countries did not respond: Turkey, Denmark, United Kingdom, and South Korea. The expertise of the participants was widespread and referred to psychological service, defense college/university, research and education departments at military academies, HR department, troop psychology, general staff operations department, and multinational joint headquarters. Functions of participants included researcher, professor, head of leadership center, director of head psychology department, troop psychologist, and lecturer at military academies or defense universities. About 54% were military employees (all with a rank of an officer) and 46% civilian employees. The descriptive statistics showed that 19% have up to ten years work experience, 43% have eleven to twenty years, and 38% have worked 21 years or more in the corresponding military organization. The participants' influence on the organizational culture addressed education, training of military personnel, research, leadership, and consultancy. On a 10-point scale (1 = *very low* to 10 = *very high*), the mean value in reference to their influence on the organizational culture is 4.80, which is considered as relatively low. The responses from the participants regarding a list of values and/or virtues being in practical use as a binding behavioral directive in their organization showed that from the 39 participants representing 15 nations, 81% confirmed that their organization uses a classification of values and/or virtues. The remaining 19% reported that they do not use a dedicated list of values and virtues. Specifically, the Indian psychological service, the Swedish Defence University, and Indonesian Defense University answered the question with a "no." In Belgium the topic is handled heterogeneously. A general list for all military people and civilians working for the ministry of defense does exist. However, employees from the Belgian Armed Forces indicated that a classification for their specific organizational unit is only partly in use. Belgian organizational units with no specific list in use included Strategy Department, HR in Ministry of Belgium Defense, and the Directorate of Material Resources within Belgian Air Forces.

Table 6 provides an overview of the outcome for military organizations where a list of values and/or virtues exists. In the last row with the title "values/virtues?" the listed expressions by the participants were categorized in values or virtues by the author of this study based on the standardized definitions as applied in the psycholexical approach (see section 6.1.1).

In the United States, each service (Army, Navy, Air Force, each police department) has its own classification. In some of the other countries, the classification of values and virtues is common for all military people and civilians working for the Ministry of Defense (e.g., Belgium). In Canada, there is a general list of values in use, which is dedicated to the four Canadian Armed Forces Values. A further list of binding values and virtues is closely linked to the levels of command (officers, NCOs, and soldiers). Austria Armed Forces indicated no determined number in their list, adapting their list of values and virtues to the situation and conveying values and virtues in training courses on subjects of political education and common values. In principle, the existing lists collected in this descriptive study cover common categories of values and virtues. The expressions that were referred to the most are loyalty, integrity, and courage. If we consider that the profession of a soldier is the same in all countries worldwide, it is not surprising to find many categories that overlap in content, even though they are not covering the same terminology.

Finally, the participants were asked to identify the overall goal in making use of the classification in the organization. Most of the participants attributed the prime objective to leadership and management, training, education, morale and culture, guiding behavior, and understanding priorities of the organization. The outcome of the inquiry also confirmed that 91% of the participants answered the question with a "yes," when asked if they believe that classification of values and/or virtues could have a positive impact on the daily military life. One of the comments brought it to the point: "Education of officers is based upon these values."

In contrast, there were also critical comments, addressing the importance of applying those values and virtues as propagated by a classification, in the sense of "Values and virtues must be lived and put into practice." An additional comment was as follows: "The fact that there are classified values to guide behavior is a good thing. However, they are taken for granted and rarely deeply discussed. For values to be meaningful, they need to be internalized."

Furthermore, the question about the official name of the classification revealed that all the mentioned names as indicated in Tab. 6 were referring explicitly to values and not to virtues. The descriptive statistics showed that 65% of the participants, however, were of the opinion that their classification combined value- and virtue-describing terms. The post-hoc categorization of the expressions into values or virtues according to standardized definitions[17] showed a ratio of 41 value-describing expressions (corresponding with the category of

---

17 The standard for the categorization was the definition by De Raad and Van Oudenhoven (2008) for values and by De Raad and Van Oudenhoven (2011) for virtues.

Tab. 6: *Overview of the core values and virtues as collected in the international study on values and virtues in military organizations*

	Australia	Austria	Belgium	Canada	Estonia
Number of values and virtues	6	no determined number	7	5	6
Classification of core values and virtues	innovation, professionalism, teamwork, courage, integrity, loyalty.	human life, physical and psychical integrity, freedom, community, fairness, justice, peace, division of powers, political co-determination, fulfillment of duty and mission, obedience of the laws, discipline, selflessness, respect, courage, comradeship, loyalty, integrity.	engagement, honor, respect, sense of service, courage, integrity, loyalty.	excellence, stewardship, courage, integrity, loyalty.	honesty, cooperation, competence, courage, loyalty, openness.
Name of list	The Defense values	-	Values of Belgium defense/les valeurs de la Défense	Canadian Armed Forces (CAF) Values	Defense Forces Values
Values or virtues?[1]	3 values & 3 virtues	-	4 values & 3 virtues	2 values & 3 virtues	2 values & 4 virtues

Note. [1]The categorization in values or virtues is made according to the definitions of De Raad & Van Oudenhoven (2009) for values and De Raad & Van Oudenhoven (2011) for virtues. The corresponding expressions are marked with colors blue (= categorized as a value) and green (= categorized as a virtue). As Tab. 6 shows, the existing classifications contained between 3 and 14 value or virtue categories.

Research on military values and virtues 101

Germany	France	The Netherlands	Norway	Singapore	South Africa	USA (Army)
8	14	3	3	8	6	7
protection of human dignity, role model in fulfillment of duty, mutual trust, ideal fostering of collaborative leadership, law-binding and responsible acting, to enforce and control orders adequately, to take care and look after the others, to inform and communicate clearly and be always approachable.	dedication, discipline, efficiency, honor, initiative, neutrality, respect, will, adaptation, competence, loyalty, modesty, moral strength, openness.	dedication, courage, resilience.	respect, responsibility, courage.	care for soldiers, discipline, ethics, fighting spirit, leadership, professionalism, safety, loyalty to country.	equal rights, honor, respect, transparency, integrity, loyalty.	duty, honor, respect, selfless service, integrity, loyalty, personal courage.
Grundsätze der Inneren Führung	Soldier's code	Core values	The Norwegian Armed Forces	8 SAF Core Value	Military Values	7 Army Values
4 values & 4 virtues	8 values & 6 virtues	1 value & 2 virtues	2 values & 1 virtue	7 values & 1 virtue	4 values & 2 virtues	4 values & 3 virtues

The data actually shows marked differences from country to country. Consequently, contents of value and virtue categories vary, which likely reflects the difference in culture, history, tradition and legal context. Additionally, the difference in core values and virtues may covariate with the type of military organization, organized in a professional (e.g., US) vs. conscript (e.g., Austria, Switzerland) army system.

values) vs. 32 virtue-describing expressions (falling into the category of virtues). This illuminates the fact that in most military organizations the view of values and virtues is not consciously determined by the distinction between the two concepts. Some important expressions, e.g., loyalty, integrity, or courage, are named and listed in some classifications as a value, although they are in fact virtues. In most cases, the term "virtue" seems to be closest to what militaries actually mean to say, when they define and discuss values.

The last questions explored whether soldiers in training vs. soldiers in combat deployment have the same view regarding what values and virtues have the highest importance. It became obvious that 65% of the participants answered "yes," indicating that the values and virtues should remain constant irrespective of the circumstances of deployment or training for operational readiness. One participant commented: "The values have to be respected in Belgium but also in deployment. I'm currently deployed abroad for two months and I can notice that these values remain of utmost importance."

In summary, the following findings resulted from this study:

- A majority (78% of responses, corresponding to 12 out of 15 countries) makes use of a specific classification of values and virtues.
- The overview of the classifications resulting from the different nations (Tab. 6) exemplifies the fact that values and virtues are described in typical country- and culture-specific ways and reflect standards in behavior. In principle, there is a degree of variety of expressions describing values and virtues, with dependencies on the function and organizational unit of the interviewed persons. Many categories of values and virtues overlap in content, even though they are expressed by different terminologies.
- The military's view on values and virtues is not consciously differentiating between values and virtues. In effect, the military reference mostly occurs in the sense of values, while implying equivalently the meaning of virtues.
- The military's view on values and virtues is not consciously differentiating between values and virtues. In effect, the military reference mostly occurs in the sense of values, while implying equivalently the meaning of virtues.
- One further outcome of this inquiry was to show that there is no corresponding research underway, conducting data analysis on the structure of military values and virtues in the international context. Although this international perspective is not exhaustive, it provided essential information, pointing to potential further research. One comment described it as follows: "The concept of values and virtues may be the core of our actions but our efforts in fundamental studies remain limited."

- The various qualitative comments of the participants confirmed that the prime concern is about the implementation of core values and virtues into the military practice. This implies that a list of core values and virtues must not only be a statement of good will and intent. The content of a classification needs to be applied within the organizational culture, i.e., by the executive military leaders acting accordingly, functioning as a role model for their employees.

4.3 Values and virtues in the Swiss Armed Forces

As recorded in the international study, most armies have more or less binding lists of core values and virtues, which are often based on traditions. The Swiss Armed Forces renounces the explicit formulation of a classification on values and virtues (see the Swiss Federal Report on Military Ethics, Swiss Armed Forces, 2010). This does not mean, however, that there is no binding base of military values and virtues. The guidelines relating values and virtues of the Swiss Federal Constitution (Swiss Confederation, 1999) and the military Service Regulations (Swiss Armed Forces, 1994) are mandatory for every soldier as representative of the executive branch. In addition, the topic of values and virtues is discussed in various military training courses within the Swiss Armed Forces. Accordingly, soldiers are ultimately bound to value these values and virtues and to comply with them. However, the renouncement of an explicit list of core values allows each military commander to set his own accents within the framework of the value and virtue base of the Swiss Armed Forces. A central position for Swiss military ethics has the value of human dignity, which is regarded as the highest value (Baumann, 2007). The Service Regulations of the Swiss Armed Forces integrate this central position in the corresponding guidelines. Thus, every member of the Swiss military organization has the duty to respect human dignity, and the military superiors are obligated not to give any orders that would violate human dignity (cf. Service Regulations 04, Swiss Armed Forces, 1994, Ziff. 77). Consequently, the Swiss Armed Forces expects its members to behave in accordance with values such as human dignity, loyalty, and responsibility and can therefore be seen as a value-oriented organization (Proyer et al., 2012). Accordingly, the doctrine of the Swiss Armed Forces (Service Regulations 04, Swiss Armed Forces, 1994) emphasizes the importance of personal values and virtues in successful leadership and military training (Annen et al., 2004). Furthermore, the official Swiss Federal Report on Military Ethics (Swiss Armed Forces, 2010) attributes increasing relevance to values and virtues. It is an essential identity of the Swiss Armed Forces to foster character development and understand the values and virtues of military persons.

There is some evidence that military experience affects political attitudes and values (French & Ernest, 1955; Jennings & Markus, 1977) and has a long-lasting influence on individual-level characteristics (Jackson et al., 2012). The primary training goals of the Swiss Armed Forces place main efforts to strengthen and influence values and virtues (Annen et al., 2004). Accordingly, these pedagogical directives make it a mandate to understand the predominant core values and virtues of the Swiss Armed Forces. More specifically, it requires a definition of those military core values and virtues that may be incorporated into the training of soldiers in the Swiss Armed Forces. As mentioned in section 1.2, values and virtues are the fundamentals to military education. The Swiss Armed Forces, with its militia system of a compulsory military service therefore is obliged, to convey the goals and content of the military education to the military cadre and civilian society with full transparency. If one understands itself as a value-based organization, one cannot ignore the need to verify the relevant contents of Swiss military values and virtues on a regular basis. Nor can it be credible to speak of military education without attempting to verify the success of this influence on the values, virtues, and behavior of the military personnel.

Accordingly, it is important to define a common classification of military values and virtues[18] within the Swiss Armed Forces. Currently, there is no generally accepted classification of military values and virtues in practice in the Swiss Armed Forces. A few approaches of values taxonomies exist (e.g., ten military pedagogical values, guiding values for military education according to Service Regulations 04); however, until now they have been developed solely based on directives, norms, or military scientific theories. Therefore, it is of great importance to establish an evidence-based classification that is systematically and empirically developed, a new and comprehensive catalog of military values and virtues that represents the culture of the Swiss Armed Forces.

5 Research questions and aim of this present thesis

The broad literature search and the summary review above confirmed that much research projects have addressed the subject of values and virtues as a part of military psychology. However, none of this research delivered a comprehensive description of the military culture. It therefore became clear that the scope of this thesis was required to extend empirical research on military values and virtues

---

18 This includes two lists, referring separately to military values and virtues within the Swiss Armed Forces.

and to better understand their structure, their relations to similar concepts and to relevant outcome variables. Accordingly, the focus was shifted to a number of open questions, specifically in reference to the psycholexical-based identification of the Swiss military values and virtues; their underlying factorial structure; their correlations with universal values, personality, and factors of character strengths; and the criterion validity of the identified military value and virtue factors.

With the theoretical background and the identified open questions in mind, data was captured and analyzed as part of three empirical studies, and a preliminary study conducted upfront. The following outline describes the specific research questions and provides an insight into the concrete steps of the Pre-study and Study I, II, and III.

Research question 1: Establishing a psycholexical derived catalog of Swiss military values and virtues
The trigger to conduct this empirical research approach was primarily given by the Swiss Report on Military Ethics (Swiss Armed Forces, 2010), which expressed a strong need in conducting research on military values and virtues, well aligned with the culture of the Swiss military organization. Although there are various written directives regarding values and virtues as typically being conveyed in training sessions of Swiss soldiers on all hierarchical levels, there is no official catalog available as a reference to the core values and virtues of the Swiss Armed Forces. If one proceeds with the viewpoint of military education (Annen et al., 2010), one has to define which values and virtues shall be conveyed to the Swiss soldiers as part of military education. As De Raad and Van Oudenhoven (2011) assumed, different cultures correlate with distinct sets of values and virtues and stated that the psycholexical procedure is a successful method to assess the cultural differences. A number of studies investigated the question of how many basic values and virtues can be distinguished through applying the psycholexical approach in a variety of cultures (Aavik & Allik, 2002; De Raad & Van Oudenhoven, 2008; Morales-Vives et al., 2012; Morales-Vives et al., 2014). In contrast, within the military domain, there has not been a systematic psycholexical consideration to shed light on the question as to which values and virtues should be listed as most important.

Taking the assumption that the military organization reflects a specific culture (Apelt, 2010), it was of particular interest to assess the military-specific values and virtues. Consequently, the following research question is defined:

<u>Research question 1</u>: Which military values and virtues can be sourced by psycholexical search from the Swiss Armed Forces' documentation and are rated as highly relevant by the executive military leaders?

To answer this question, this thesis aligned the psycholexical approach with the specific culture of the military environment and followed the methodological principles as illustrated in the Dutch studies of De Raad and Van Oudenhoven on values (2008) and virtues (2011). Therefore, the aim of the Pre-study was to develop a catalog of military value and virtue descriptors reflecting the relevance within the Swiss Armed Forces on a psycholexical-oriented basis.

Using a psycholexical search of military documentation (e.g., Swiss military Service Regulations 2004), combined with the ratings of military psychologists and top military leaders, a catalog of the military values and military virtues (MVVC) was developed. Subsequently, this catalog of military values and virtues was used in all three empirical studies to gain insight into the subjective ratings of different military subgroups within the Swiss military organization.

Research questions 2 and 3: Assessing the structure of military values and virtues and the correlations with related constructs

A related question concerned the structure of military values and virtues. The Swiss Armed Forces does not have a representative classification of core values and virtues, as, for instance, it applies to US Army doctrine in Field Manual 22–100 with loyalty, duty, respect, selfless service, honor, integrity, and personal courage (LDRSHIP; U.S. Department of the Army, 1999, p. B-2) or to the Canadian Armed Forces with integrity, loyalty, courage, stewardship, and excellence (Canadian Armed Forces Website, 2017).

An intercultural comparison between the value and virtue classifications within the international descriptive study of various military organizations (see section 4.2) allowed to conclude that the content, terminology, and understanding of the core values and virtues are subject to cultural variation. Therefore, an ultimate aim of this present thesis was to assess the factorial structure of the Swiss military values and virtues and to derive the core values and virtues. The structure of military values and virtues was analyzed separately, focusing on military values in Study I and on military virtues in Study II.

First, within research questions 2a and 2b the focus was on the military values:

<u>Research question 2a</u>: What are the factors of Swiss military values?

This research question corresponded with the question of how the military value factors relate to other correlates. Psycholexical studies on universal values (e.g., Aavik & Allik, 2002; Morales-Vives et al., 2012) have all included personality measures to examine the relations to the statistical value factors. Up to now, only measures of universal values and not military values have been linked to personality characteristics. Nor have any studies included both measures of universal and military values and looked at the correlation between these measures.

> Research question 2b: How do the military value factors relate to universal values factors of personality?

In Study I, the MVVC as developed in the Pre-study was administered to Swiss career officers and career NCOs. The aim of this study was (a) to analyze the structure of Swiss military values within this sample, and (b) to examine as to how the factorial structure of the military values corresponds to the five universal value factors of Renner (2003b), and the five factors of personality. Mapping the military value factors with the universal values and personality characteristics made it possible to assess the distinct nature of the identified military value factors.

Second, the question of the structure of military virtues has not yet been researched and was addressed in research questions 3a and 3b. The outcome of the empirical research regarding the factorial structure of Swiss military virtues provided a solid framework to define the core virtues of the Swiss Armed Forces. In analogy to research question 2a, this led to the research question 3a, with specific reference to military virtues:

> Research question 3a: What are the factors of Swiss military virtues?

The development of the VIA classification by Peterson and Seligman (2004) was part of the beginning of positive psychology and its initial aim. As Matthews (2009) pointed out, the military environment is a natural home for the concept of positive psychology. It ultimately raised the question about the factorial structure of the military virtues and how it relates to the second-order factors of the 24 character strengths, assessed by the VIA-IS (Peterson et al., 2005). It was hypothesized that there would be similarities and partial overlap as well as differences. This added verification as to how the outcome of the military virtue factors compare with the factorial structure of the VIA-IS, which has neither been discussed in the literature nor examined empirically at a detailed level.

> Research question 3b: How do the military virtue factors relate to the five second-order factors of character strengths in the Values in Action Inventory of Strengths (VIA-IS)?

In answering research questions 3a and 3b, as part of Study II, military virtues among future militia cadre was analyzed in reference to the corresponding structure and the relation to the VIA-IS. The objective was (a) to structure the Swiss military virtues, and (b) to analyze as to how the military virtue structure corresponds with the structure of the five factors of character strengths. The latter was a prerequisite to interpret the outcome of the military virtue factors as distinct factors and to compare the factorial structures of the two models.

It is important to draw attention to the fact that the objectives between Study I and Study II were very similar, with the distinction of correlating the military values with five factors of universal values and personality (Study I), and the military virtues with five factors of character strengths (Study II).

Research question 4: Investigating the criterion validity of the military value and military virtue factors
Previous research on the outcome of military values and virtues primarily addressed variables of success within the context of leadership and operational training. This thesis widened the perspective by examining OCB and MTL as criteria variables. Empirical results showed that personal values influence motivation in the form of organizational behavior (Ajzen, 1991) and voluntary work (Omoto & Snyder, 1995). Consequently, military persons who associate well with the military values and virtues were expected to show motivation to pursue a career as a militia cadre and to behave in line with OCB.

> Research question 4: Do the military value and virtue factors predict organizational citizenship behavior (OCB) and motivation to lead (MTL) as criteria variables?

The Study III concerned the relationship of military values and virtues with OCB and MTL and assessed the incremental validity of military values and virtues extending the scope beyond universal values. The main aim was (a) to assess the ratings by Swiss soldiers regarding the descriptive terms of the 25 military values and 42 military virtues via the MVVC; (b) to analyze as to how the universal values, military values, and military virtue relate to OCB and MTL; and thus (c) to validate the military values and military virtues, determining whether there is incremental validity beyond the universal values.

In summary, this thesis aimed at contributing to the progress in current research on military values and virtues. The prime focus was to identify descriptors of military values and virtues based on principles of the psycholexical approach; to determine the inherent structure of the military values and virtues; and to assess how the factorial structure correlates with universal values, the Big Five of personality, and factors of character strengths. This provided evidence regarding the extent to which the military environment differs from the civilian context. Furthermore, the research examined to what extent military values and virtues allow a prediction regarding OCB and MTL.

6 Procedure

In this last section of the theoretical background, two issues need to be addressed, regarding the applied methodology in research design: It concerns in section 6.1

the development of a psycholexically derived inventory of 25 military values and 42 military virtues (MVVC), and in section 6.2 the hierarchical top-down procedure in collecting data assessing military values and virtues.

6.1 Pre-study: Development of the MVVC

In a preliminary step, a psycholexically derived list of 25 military values and 42 military virtues was created. The selection and practical validation of the value and virtue descriptors was conducted in three stages, in line with the psycholexical studies as reviewed above in section 3.4.

6.1.1 Psycholexical-oriented analysis

To reach the overall goal of a full list of military values and virtues, a psycholexical-based search of relevant military documentation was conducted in a first step. As described in section 3.4, a lexical approach generally uses a dictionary of words to identify descriptors of human behavior. Accordingly, the existing written military guidelines were analyzed, which included military directives to enforce binding behavior for every military person as a member of the Swiss Armed Forces. Every important normative behavior and principle of military life is represented in these normative documents. Hence, this approach does not hinge upon the definition of lexical words and synonyms, but rather makes reference to behavioral mandates[19]. In specific, the Swiss Federal Report on Military Ethics (Swiss Armed Forces, 2010), the Service Regulations of the Swiss Armed Forces (Swiss Armed Forces, 1994), and several sources of military literature on values and virtues in the Swiss Armed Forces (e.g., Annen et al., 2004) were analyzed. Table 7 provides the overview of the military documentation and examples of various military value- and virtue-describing expressions that were analyzed to establish a psycholexical-based procedure.

As seen from Tab. 7, the psycholexical content includes the existing guidelines on Swiss military values and virtues, in spite of the possible degree of redundancy of the expressions. The aim was to collect a cross-organizational coverage

---

19 From a methodology viewpoint, the descriptions of behavioral mandates as used in the military documentation are presented differently than in a standardized lexicon: There were no glossaries of expressions, no use of synonyms, and no explicit differentiation of adjectives and nouns. The descriptions sourced from the military documentation are already preselected terms that are considered authorized and in effect from a normative military perspective. Accordingly, the terms "psycholexical-based" or "psycholexical-oriented" are in use within this framework of research.

Tab. 7: *List of main sources on Swiss military values and virtues used for the psycholexical search (adapted from the Swiss Federal Report on Military Ethics, 2010, p. 24)*

Sources of military documentation	Extracted military value- and virtue-describing terms
Reglement Grundschulung der Schweizer Armee [Regulations for basic military training of the Swiss Armed Forces]	*Soldier level:* discipline, persistence, competence, courage, self-confidence NCOs *and officers level:* role model, charity, responsibility, initiative
Behelf für Offiziere des Zentralen Offizierslehrgangs [Instructions for officers in training]	loyalty, sense of duty, sense of responsibility, respect, solidarity, integrity, boldness, honor
Werte der Infanterie [Guiding values of Infantry]	loyalty, trust, will
Werte der Grenadiere [Guiding values of Grenadiers]	honor, modesty, union
Vorgaben der Infanterieoffiziersschule [Regulations of the Infantry Officer School]	persistence, discipline, honesty, loyalty, courage, punctuality, willingness to perform, esprit de corps, sense of responsibility, respect
Werte der militärischen Erziehung gemäss Dienstreglement 04, Ziff. 33 [Guiding values for military education according to Service Regulations 2004, Ziff. 33]	comradeship, trust, discipline, hierarchy, teamwork, persistence, loyalty
Zehn Werte der militärischen Erziehung gemäss Annen et al. (2004, p. 106–110) [Ten guiding values for military education according to Annen et al. (2004)]	human dignity, trust, personal responsibility, initiative, integrity, comradeship, loyalty, personal courage, performance of duty, unselfishness
Werte einer "integrierten Militärethik" gemäss Baumann (2007) [Values of a "integrated military ethics" according to (2007)]	human dignity, obedience of the law, sense of responsibility, moral power of judgment, performance of duty, courage, discipline, obedience, comradeship, loyalty, integrity

of values and virtues, reflecting the culture of the Swiss Armed Forces. It is important to mention that all the sources are based on normative guidelines and as such are not yet empirically verified.

To scan these documents for expressions used to describe military values and virtues, the first author of this study teamed with an experienced military lead commander. The psycholexical search for the military value- and virtue-describing terms was made together by both persons and was agreed on by consensus. The intent was (a) to select expressions that refer to desirable and positive characteristics or behavior of a military person, and (b) to give priority to nouns over adjectives or verbs (i.e., four adjectives and verbs were completely excluded; e.g. having overview). According to Aavik and Allik (2002), nouns are preferable for expressing values because people usually think of values in noun form. Based on these guidelines, 90 expressions were extracted. To obtain an exhaustive list of military values and virtues, a conservative process was followed, excluding seven terms that do not express human behavior or thoughts (e.g., trimness, smartness; according to the procedure in the study of Morales-Vives et al., 2012).

6.1.2 Categorization in values and virtues

To ensure that values and virtues were considered separately, the second step was the following, in which eleven judges rated the 90 value- and virtue-descriptive words. These judges were licensed psychologists with an affiliation to the military culture. All were experienced assessors in the selection of career officers and NCOs. They rated the extent to which those terms could be considered as a value within the Swiss military culture and were instructed to indicate for each of the 90 expressions their individual rating according to the clear definition that a value is "a relatively enduring characteristic of individuals that reflects what is important to them and that guides them in their behaviors and their decisions" (De Raad & Van Oudenhoven, 2008, p. 85–86). Analogue to the procedure of De Raad and Van Oudenhoven (2008) a 4-point scale was used with 1 (*is clearly not a value*), 2 (*is probably not a value*), 3 (*is probably a value*), and 4 (*is very clearly a value*). For virtues the following definition was used: "Virtue is broadly defined as a moral trait, indicating what one should be or do or show, demonstrate, respect, etc., depending on the form of the term in question" (De Raad & Van Oudenhoven, 2011, p. 45), using a 4-point scale with 1 (*is clearly not a virtue*), 2 (*is probably not a virtue*), 3 (*is probably a virtue*), and 4 (*is very clearly a virtue*).

For each term, a sum score was calculated over the eleven subjects, which was an index of value- or virtue-descriptiveness. For the values the sum was ranging from 13.00 to 44.00. The mean of the sum scores was 29.00 ($SD = 7.70$). According to the methods of the Dutch value study (De Raad & Van Oudenhoven, 2008), all terms with a sum-score of 33.00 or higher were identified as value descriptors (sum-scores below 33.00 meant that judges on average did not regard these expressions as a value). The same procedure was applied to assess the virtue-descriptiveness. Thirteen terms did not meet the criteria for neither being a value nor a virtue and were skipped accordingly.

To control and verify the accuracy of the ratings, ten distracting and nonrelevant terms were inserted, e.g., shoe size, marital status, and impatience, which are definitely not defined as values. All eleven judges recognized the "false" values or virtues and rated these terms with "1" or, a maximum of, "2" (which was the case for multilingualism and impatience by two raters). The resulting list consisted of 25 military values such as human dignity, role model, peace, and hierarchy, and 42 military virtues such as punctuality, courage, integrity, and loyalty.

6.1.3 Expert interview with high executive military leaders

In a third stage, 22 high ranking leaders of the Swiss Armed Forces were consulted and subjective ratings are collected, to ensure that the 25 targeted military value-describing terms and 42 military virtue-describing expressions are relevant from a military perspective. The 22 individual military leaders (from a population of 35 generals) were identified in accordance with their military rank (Brigadier General and higher) and function (e.g., commander of a training unit), in representation of the respective influence to promote the corresponding values and virtues within "top-down hierarchy" of the Swiss Armed Forces. Full support from all of these top executive leaders was received, including the highest military position with the chief of the Swiss Armed Forces (Lieutenant General).

Military values

The generals were asked to rate each value to the extent that it should normatively provide orientation for the military person in his or her everyday military decisions and actions. A Likert scale was used with 1 (*military persons should not orient themself by this value*), 2 (*military personnel should orient itself often by*), 3 (*military personnel should orient itself when always possible by*), and 4 (*military persons should orient themself at all costs by this value*). Mean inter-rater correlation (Pearson correlations) was .63, with the highest correlation between two raters of .70 and the lowest of .54. Descriptive statistics of the ratings with the military experts are shown in Tab. 8.

As seen in Tab. 8, human dignity ($M = 3.82$, $SD = 0.50$) and honesty ($M = 3.82$, $SD = 0.40$) were rated at the highest mean, and autonomy the lowest ($M = 2.45$, $SD = 0.18$). Also noticeable was a slight ceiling effect in the data. Average mean of the 25 items over the military raters was 3.21, with a median of 3.14, indicating that the selected value descriptors were considered important by the military experts. There was a significant difference between human dignity ($M = 3.82$, $SD = 0.50$) and autonomy ($M = 2.45$, $SD = 0.18$) ($t(20) = 5.73, p < .001$, $d = 0.35$), implying that the expert could sufficiently differentiate between the

Tab. 8: *Descriptive statistics of 25 military value descriptors from the perspective of the Swiss military Generals*

Military value descriptors		Descriptive statistics						
German term	English term	M	Med	SD	S	K	Min	Max
Menschenwürde	human dignity	3.82	4.00	0.50	-2.91	8.43	2	4
Respekt	respect	3.55	4.00	0.51	-0.20	-2.17	3	4
Kameradschaft	comradeship	3.23	3.00	0.53	0.26	0.14	2	4
Solidarität	solidarity	3.00	3.00	0.54	0.00	1.18	2	4
Vertrauen	trust	3.41	3.00	0.50	0.40	-2.04	3	4
Pflichterfüllung	performance of duty	3.59	4.00	0.59	-1.15	0.51	2	4
Auftragserfüllung	performance of mission	3.64	4.00	0.49	-0.61	-1.80	3	4
Gehorsam	obedience	2.95	3.00	0.58	-0.01	0.51	2	4
Einordnung	integration	2.82	3.00	0.50	-0.41	0.75	2	4
Hierarchie	hierarchy	2.86	3.00	0.56	-0.07	0.46	2	4
Sicherheit	security	3.36	3.00	0.66	-0.55	-0.53	2	4
Frieden	peace	3.00	3.00	0.82	0.00	-1.48	2	4
Freiheit	freedom	2.95	3.00	0.90	-0.34	-0.76	1	4
Gerechtigkeit	justice	3.18	3.00	0.59	-0.03	0.01	2	4
Rechtstreue	obedience of the laws	3.36	3.00	0.58	-0.21	-0.62	2	4
Vorbildlichkeit	role model	3.73	4.00	0.46	-1.10	-0.89	3	4
Ehre	honor	2.68	3.00	0.65	-0.76	1.05	1	4
Multikulturalität	multiculturalism	2.73	3.00	0.63	0.27	-0.46	2	4
Korpsgeist	esprit de corps	3.05	3.00	0.63	-0.04	-0.37	2	4
Achtung des Nächsten	respect of the next ones	3.68	4.00	0.48	-0.84	-1.44	3	4
Zusammenarbeit	teamwork	3.05	3.00	0.58	0.01	0.51	2	4
Ehrlichkeit	honesty	3.82	4.00	0.40	-1.77	1.25	3	4
Fairness	fairness	3.14	3.00	0.64	-0.11	-0.32	2	4
Zusammenhalt	coherence	3.09	3.00	0.43	0.64	3.17	2	4
Autonomie	autonomy	2.45	2.00	0.67	0.18	0.10	1	4

Note. N = 22. M = mean, Med = median, SD = standard deviation, S = skewness, K = kurtosis, Min = minimum, Max = maximum.

value descriptors. The inter-rater reliability for the 25 military values over the 22 raters was .75. In accordance to the procedure of De Raad and Van Oudenhoven (2008), all those terms were retained as value descriptors that had a mean (across the 22 military experts) of 2.00 and higher. This is the set of terms on average considered to be important in the military context, on the given rating scale from 1 to 4. Since no term had a mean under 2.00, it was concluded that these value descriptors were relevant from a military point of view, and thus decided to retain the full list of 25 military values.

Military virtues

The generals were asked to rate each virtue in how far military personnel should ideally possess the appropriate characteristic. A Likert scale was used, ranging from 1 (*negligible*), 2 (*important*), 3 (*very important*), to 4 (*essential*).

The mean inter-rater correlation was .74, with the highest correlation between two raters of .78 and the lowest of .41. Descriptive statistics of the ratings with the military experts are shown in Tab. 9.

As seen in Tab. 9, personal responsibility ($M = 3.77$, $SD = 0.43$) and sense of duty ($M = 3.55$, $SD = 0.51$) were rated at the highest mean, and charity the lowest ($M = 2.01$, $SD = 0.61$). As in the answers on military values a slight ceiling effect in the data occurred. Average mean of the 42 items over the military raters was 2.97, indicating that the selected virtue descriptors were considered important by the military experts. The inter-rater reliability for the 42 military virtues over the 22 raters was .94. Since no term had a mean under 2.00, it was concluded that these value descriptors were relevant from a military point of view, and thus decided to retain the full list of 42 military virtues.

In summary, the military values and virtues identified by psycholexical search were all confirmed to be of high relevance from a military perspective. The analysis conducted within this stage of expert interview indicated that the executive military leaders agreed with the selected value and virtue descriptors. This preliminary study established a list of 25 military value descriptors and one of 42 military virtue descriptors, named as the MVVC[20].

6.2 The MVVC applied in a hierarchical top-down data collection

Within the Pre-study, the two lists of the MVVC were established by means of a psycholexical search, a categorization into values and virtue with ratings by military psychologists and a practical validation interview with top executive military leaders as military experts.

---

[20] The name "catalog" indicates that it is a compilation (in form of a list) of the most important military values and virtues. It is not a validated psychometric instrument yet to measure the value and virtue factors.

Tab. 9: *Descriptive statistics of 42 military virtue descriptors from the perspective of the Swiss military Generals*

Military virtue descriptors		Descriptive statistics						
German term	English term	M	Med	SD	S	K	Min	Max
Rücksichtnahme	consideration	2.50	2.00	0.60	0.74	-0.31	2	4
Fürsorge	welfare	3.00	3.00	0.76	0.00	-1.15	2	4
Treue	faithfulness	2.91	3.00	0.68	0.11	-0.65	2	4
Eigenverantwortung	personal responsibility	3.77	4.00	0.43	-1.40	-0.06	3	4
Verbindlichkeit	liability	3.23	3.00	0.75	-0.41	-1.04	2	4
Disziplin	discipline	3.41	3.50	0.67	-0.70	-0.43	2	4
Initiative	initiative	3.05	3.00	0.79	-0.08	-1.32	2	4
Mässigung	moderation	2.68	3.00	0.65	0.40	-0.54	2	4
Ehrenhaftigkeit	sense of honor	2.77	3.00	0.69	-0.65	1.18	1	4
Wille	will	3.32	3.00	0.57	-0.05	-0.51	2	4
Wohltätigkeit	charity	2.01	2.00	0.61	0.03	0.03	1	4
Leistungswille	motivation	3.50	4.00	0.60	-0.74	-0.31	2	4
Staatsbürgerliche Toleranz	political tolerance	2.55	2.00	0.91	0.27	-0.69	1	4
Verlässlichkeit	reliability	3.36	3.00	0.66	-0.55	-0.53	2	4
Integrität	integrity	3.45	3.50	0.60	-0.55	-0.52	2	4
Authentizität	authenticity	2.91	3.00	0.75	0.15	-1.11	2	4
Pflichtbewusstsein	sense of duty	3.55	4.00	0.51	-0.20	-2.17	3	4
Gewissenhaftigkeit	conscientiousness	3.18	3.00	0.66	-0.21	-0.55	2	4
Glaubwürdigkeit	credibility	3.23	3.00	0.61	-0.14	-0.29	2	4
Selbstlosigkeit	selflessness	2.55	2.00	0.80	0.45	-0.33	1	4
Loyalität	loyalty	3.45	4.00	0.74	-1.00	-0.32	2	4
Uneigennützigkeit	unselfishness	2.59	2.00	0.73	0.85	-0.54	2	4
Tapferkeit	bravery	2.41	2.00	0.67	0.37	0.27	1	4
Mut	courage	2.73	3.00	0.63	0.27	-0.46	2	4
Selbstvertrauen	self-confidence	3.05	3.00	0.49	0.15	2.08	2	4
Bescheidenheit	modesty	2.82	3.00	0.73	0.30	-0.97	2	4
Zivilcourage	moral courage	2.86	3.00	0.83	-0.27	-0.36	1	4
Unerschrockenheit	boldness	2.27	2.00	0.63	-0.27	-0.46	1	4
Leistungsbereitschaft	willingness to perform	3.18	3.00	0.73	-0.30	-0.97	2	4
Ausdauer	endurance	2.82	3.00	0.66	0.21	-0.55	2	4
Durchhaltevermögen	persistence	3.05	3.00	0.65	-0.04	-0.37	2	4

(continued on next page)

Tab. 9: Continued

		Descriptive statistics						
Selbstständigkeit	independence	3.09	3.00	0.68	-0.11	-0.65	2	4
Lernbereitschaft	willingness to learn	3.09	3.00	0.75	-0.15	-1.11	2	4
Entscheidungsstärke	decision-making qualities	3.32	3.00	0.72	-0.57	-0.76	2	4
Moralische Urteilsfähigkeit	moral power of judgment	3.14	3.00	0.77	-0.25	-1.23	2	4
Weisheit	wisdom	2.36	2.00	0.85	0.21	-0.29	1	4
Verantwortungs-bewusstsein	sense of responsibility	3.55	4.00	0.51	-0.20	-2.17	3	4
Klugheit	prudence	2.50	2.00	0.74	1.16	-0.02	2	4
Beharrlichkeit	tenacity	2.91	3.00	0.68	0.11	-0.65	2	4
Kritikfähigkeit	ability to take criticism	2.77	3.00	0.75	-0.33	0.26	1	4
Pünktlichkeit	punctuality	2.82	3.00	0.80	0.35	-1.29	2	4
Besonnenheit	considerateness	2.91	3.00	0.75	0.15	-1.11	2	4

Note. $N = 22$. M = mean, Med = median, SD = standard deviation, S = skewness, K = kurtosis, Min = minimum, Max = maximum.

The MVVC provided the basis questionnaire within this thesis and was used for all subsequent studies. Figure 3 shows the basic structure of the MVVC as applied in the data collection.

As retrieved from Fig. 3, the MVVC consisted of two parts and aimed on collecting the subjective ratings relating the 25 military value items and 42 military virtue items. One of the concerns referred to the format of the answers of the MVVC. As identified in the Pre-study, there was a risk to have a ceiling effect within the rating scope of the military values and virtues. In order to avoid such uncontrolled answer tendencies another pretesting was conducted evaluating if the 4-point or 7-point scale makes a difference in the variance of ratings. In similar studies a 4- or 5-point scale was used (De Raad & Van Oudenhoven, 2008, 2011). This pretesting was with the objective to receive initial evidence for the data distribution in such specialized sample of military persons who are socialized to those kind of value- and virtue-describing terms. The MVVC was therefore administered to 69 career officer students at Military Academy in a within-subject design, specifically having the participants rate the same 25 military values in a 4-point and 7-point format. The results with the 69 career officer students pointed to the conclusion to choose the 7-point scale. The coefficient of variation (CV)

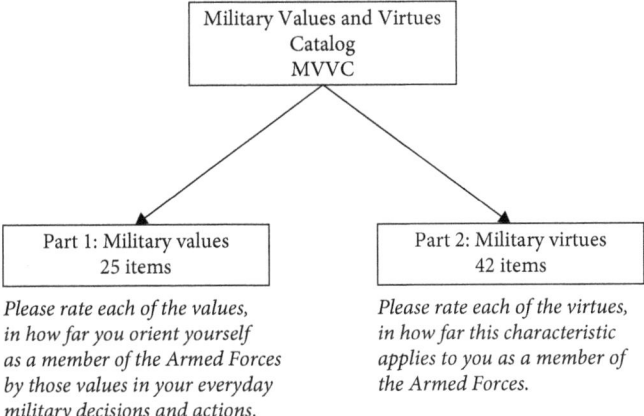

Fig. 3: Structure of the MVVC

was calculated as a normalized measure of dispersion of a probability distribution. The mean of the CV value measures was slightly higher for the 7-point scale (CV_{mean} = 21.17) than for the 4-point scale (CV_{mean} = 21.00). Additionally, we asked the participants at the end of both answer format block to what extent they could differentiate between the several categories and how much they felt at ease with the rating scale. With both respects the participants preferred the 7-point scale to the 4-point scale. As a conclusion, a 7-point scale was used in the subsequent studies for both parts of the 25 military values and the 42 military virtues.

Within the given context of the Swiss Armed Forces, data was collected in a hierarchical top-down design and sequence. The ratings of various groups of military persons are studied by assessing the ratings on military values and virtues of various groups of military persons in the sequence of executive military leaders, military professionals such as career officers and career NCOs, militia officer candidates, and militia recruits (see Fig. 1 for an overview). Accordingly, the MVVC was administered to large samples of military subgroups within the Swiss Armed Forces in Study I, II, and III to assess the factorial structure of military values and virtues.

In Study I, the ratings of the military professionals (i.e., career officers and career NCOs) were assessed. This group of military leaders propagates the desired values and virtues of the lead commanders down to the lower hierarchical levels. In Study II, the MVVC was administered to militia officer candidates; and in Study III the lowest hierarchical level of the recruits was addressed. By collecting self-ratings on the items of military values and virtues, the studies took individual differences into account. The outcome of the data analysis provided a conclusive view as to how consistent value- and virtue-based messages are communicated and conveyed from top-down military hierarchical levels.

Study I: Assessing the Structure of Military Values

7 The structure of military values and the relation to universal values and personality

7.1 Introduction

There is little verification through empirical evidence defining which values characterize the desired military culture and what factorial structure describes such a value catalog. In contrast, outside of the military domain, there has been an increasing interest in studies identifying and structuring taxonomies of universal values. Specifically, the question of how many basic values can be distinguished has been investigated through the psycholexical approach in a variety of different cultures (e.g., Aavik & Allik, 2002; De Raad & Van Oudenhoven, 2008; Morales-Vives et al., 2012; Renner, 2003b). This study is the first to empirically assess the structure of values in a military organization by means of a psycholexical and factor analytic approach. The aim was to identify Swiss military values, to define an organizing structure, and to validate the distinct nature of military value factors, as they relate to universal values and personality traits.

7.1.1 The psycholexical approach towards the structure of values

Psychological literature refers to numerous definitions and theoretical approaches to describe values. It is not surprising therefore that definitions vary. According to Hitlin and Piliavin (2004, p. 362), values can be conceived as a formation of an "internal moral compass." Also, values are usually considered as conceptions of the desirable, which influence the way people make choices and evaluations (Kluckhon, 1951; Van Deth, 1995). Similarly, Schwartz (1994, 1999) focused on the motivational power of values and defined them as desirable goals that vary in importance across situations and that guide the way social actors (e.g., individual persons such as military leaders) choose actions and evaluate people and events. Rokeach (1973b) contrasted this view with another common definition: Values are "enduring beliefs that a specific mode of conduct or end-state of existence is personally or socially preferable to an opposite or converse mode of conduct or end-state of existence" (p. 5). He was interested in a full set of values to describe an individual view and implemented two distinct lists of 18 instrumental values (describing modes of conduct as forms of behavior) and 18 terminal values (describing end-states of existence as lifetime goals). Unlike Hofstede (1980) or Schwarz (1994, 1999) who were interested in values as they manifested themselves at a sociocultural level, Rokeach studied values as they pertain to individuals. In alignment with previous studies on interpersonal differences in values, in the present study a value is understood to be "a relatively enduring characteristic of individuals that reflect what is important to them and that guides them in their behaviors and decisions" (De Raad & Van Oudenhoven, 2008, p. 85–86).

The question of how many basic values can be distinguished has been investigated within two domains of research, with little connection to each other. These domains refer to (1) Vernon and Allport (1931) and to (2) Rokeach (1973b) and Schwartz (1992, 1999). These research studies mainly followed a theoretical approach, based on a rational analysis (De Raad & Van Oudenhoven, 2008). In recent years, the conceptional framework of Schwartz (1992, 1994) and Schwartz and Bilsky (1990) has strongly influenced the research on values. By means of a facet-theoretical approach, they initially identified seven motivational domains of values and ultimately increased it to ten domains, representing a universal structure of human values. However, the drawback of the taxonomies derived through rational analysis is with the identification of the facets and their elements being subjective and dependent on the insights of the theoretician (De Raad & Hendriks, 1997).

Due to this restriction in methodology, Aavik and Allik (2002) gave preference to the psycholexical approach as a basis to develop an exhaustive and culturally sensitive list of universal values in the Estonian language. Given the application of the psycholexical approach, which traditionally has been used to classify personality traits, they initiated a new method of value research. A series of researchers followed this novel study design immediately, e.g., Renner (2003b; for the German language in Austria); Renner, Peltzer, and Phaswana (2003; for the Northern Sotho language in South Africa); Renner and Myambo (2007; for the Egyptian language); and De Raad and Van Oudenhoven (2008; for the Dutch language in the Netherlands). In subsequent studies Morales-Vives et al. (2012; for the Spanish language) and Cretu, Burcas, and Negovan (2012; for the Romanian language) conducted further psycholexical studies on values.

The psycholexical approach assumes that language will represent the most significant individual differences in daily transactions with other people (Goldberg, 1981). Accordingly, the so-called lexical approach defines the lexicon in its particular language as the basis to identify the descriptors of human behaviors. As such, the lexical approach does not depend on the accuracy of preexisting scientific concepts (De Raad, 2000). The three psycholexical studies of Aavik and

Allik (2002), Renner (2003b), and De Raad and Van Oudenhoven (2008) have predominantly influenced the focus of this current research (for an overview of the systems of values see Tab. 4).

Aavik and Allik (2002) selected 78 value descriptors from a list of 560 words, extracted from an Estonian lexicon. The expressions were given to 294 participants, who were asked to rate them according to the extent in which they provided guidance to their lives. Principal component analysis revealed six factors, labeled Self-Enhancement, Self-Realization, Hedonism, Benevolence, Broadmindedness, and Conservatism.

In Austria, Renner (2003b) made the distinction between instrumental values and terminal values. Some 383 nouns and 299 adjectives were selected from the German lexicon and provided to 456 participants who were asked to rank them in a study following the same design. Principal component analysis was performed, delivering four instrumental value factors and five terminal value factors, turning out to be equal. Renner concluded that individuals do not distinguish between instrumental and terminal values. The five core factors were labeled Profit, Balance, Salvation, Intellectualism, and Conservatism (which was the additional terminal value factor).

De Raad and Van Oudenhoven (2008) conducted the study in Dutch, beginning with 4659 expressions, which were stepwise reduced to 641 values. They administered the extracted expressions to 634 participants, concluding with eight factors, being Status and Comfort, Organization and Achievement, Competence, Love and Happiness, Benevolence, Spirituality, Family and Tradition, and Aesthetics and Erudition.

The number of extracted value terms varied between the studies (560 to 4660 words)

The number of the extracted value terms varied between the studies (560 to 4660 value expressions). This can be explained by the differently used definition and the varying procedure, and of course, the peculiarity of the language played a role. In sum, the three studies described above delivered between four and eight factors of universal values. Some of the values are replicated in different cultures (e.g., Benevolence) or correspond to equivalent factors with different labels (such as Status and Comfort by De Raad & Van Oudenhoven, 2008, and Self-Enhancement by Aavik & Allik, 2002). Other values seem specific to the culture and accordingly could not be easily replicated (e.g., Aesthetics and Erudition from the De Raad & Van Oudenhoven study). This result aligns with the assumption by De Raad and Van Oudenhoven that different cultures have distinct sets of values. Accordingly, one can conclude that the psycholexical assessment is an useful tool for revealing cultural differences in values. Taking the assumption that the military organization reflects a specific culture (Apelt, 2010) makes it of particular interest to assess the actual military values, its factorial structure and to draw conclusions on the nature of the military culture.

7.1.2 Relations of values to personality

Most of the psycholexical studies mentioned above (i.e., De Raad & Van Oudenhoven, 2008; Morales-Vives et al., 2012; Renner, 2003b) have integrated personality measures in their analyses and have addressed the relationships between value factors and Big Five trait factors. Personality traits are understood as relatively stable characteristics that determine how people think, feel, and act (Johnson, 1997). De Raad and Van Oudenhoven (2011) describe personality traits as characteristics that *dispose* a person to behavior, while values are characteristics that influence what a person finds *important* and that *guide* a person in behavior and decisions. In accordance, Roccas, Sagiv, Schwartz, and Knafo (2002) pointed out that traits are enduring dispositions and values are enduring goals. Moreover, the literature has specifically identified the Big Five factor of Agreeableness as morality-related personality traits (e.g., Strelau & Zawadzki, 2006) and concluded that they "may therefore be considered as the more typical value-laden character factors" (De Raad & Van Oudenhoven, 2008, p. 102). Several studies found significant correlations between Agreeableness and values such as Benevolence, Love, Balance, Universalism, Security, and Conformity (positively) and Status, Profit, and Comfort (negatively) (De Raad & Van Oudenhoven, 2008; Morales-Vives et al., 2011; Renner, 2003b). Furthermore, people with higher levels of Conscientiousness tend to prefer values such as Responsibility, Organization, Achievement, Conservatism, Professionalism, Family, and Tradition (Renner, 2003b; De Raad & Van Oudenhoven, 2008; Morales-Vives et al., 2011). Correlations to the other personality factors such as Extraversion and Openness to Experience were also found in the same referenced studies. Extraversion was shown to relate to Hedonism, Happiness, and Social relationships. Additionally, Openness to Experience was consistently found to correlate positively with values of Universalism and Stimulation and negatively with Security and Tradition. The trait factor Emotional Stability was consistently found to not correlate with any value factors (Morales-Vives et al., 2011). From the overall findings it was concluded that Big Five personality traits do correlate with values and do allow an indicative interpretation in accordance with the distinct meaning of a particular value descriptor. Being mentioned again, there has been no study so far that investigates the specific relationship as it applies to military values and personality.

7.1.3 Research on values in military psychology

There is growing evidence that understanding soldier performance requires addressing both cognitive and non-cognitive factors, i.e., values and personality (Böhm, 2008). This specifically is true in the military environment, where decisions are often made under challenging conditions, including sleep deprivation, and fear of death or severe bodily harm (Kornguth, Steinberg, & Matthews, 2010). Consequently, positive characteristics of personality like character strengths, virtues, and values are recognized as critical to military leadership (Matthews et al., 2006b).

In spite of the emphasis in priority, little psychological theory or empirical evidence exists which underlines the importance of values within the military context (Grojean & Thomas, 2006). Corresponding studies investigating the extent to which the military culture shapes the values of military leaders have been conducted in the US military (e.g., Bachman, Sigelman, & Diamond, 1987), as well as in Germany (Jackson, Thoemmes, Jonkmann, Lüdtke, & Trautwein, 2012), South Africa (Franke & Heinecken, 2001), Australia (Mc Allister & Smith, 1989; Mc Allister, 1995), and Japan (Inagaki, 1975). Recent research focuses on the fundamental question of how to ensure a desired military culture by means of selection and socialization (for an overview, see Jackson et al., 2012). Thus, there has been little research about the structure of military values which could help to describe the value culture of a military organization and to better understand the influence of cultural components on character development of recruits and on external variables such as leadership performance and job satisfaction.

Schumm, Gade, and Bell (2003) explored the structure of military values in the US Army. By using various measures of professional military values they assessed the structure and how the military value factors related to military outcome variables such as retention, morale, satisfaction with the quality of Army life, overall job satisfaction, and wartime preparedness. They administered a questionnaire of 15 value-related items to 7,860 soldiers in the US Army. The result of the factor analysis based on the collected ratings was a four-factor solution, i.e., (1) Military Dedication (loyalty to the US Army, to the military unit, the nation, and to a willingness to risk one's life for the military organization); (2) Integrity (being honest and doing what is right); (3) Military Bearing (maintaining a military appearance, high moral standards on- and off-duty, and showing proper respect and military courtesy to others); and (4) Job Commitment (job dedication, personal drive to succeed, and being committed to working as a team). Military Dedication exhibited the largest correlation with outcome variables such as individual morale, retention intentions, and deployment readiness.

Furthermore, high correlation was found between military values and selected demographic variables (i.e., rank, years of service, age). The authors therefore suggested the inclusion of multivariate analysis based on value factors, demographic variables, and outcome variables. The five scales of military values derived in this study showed very high internal consistency as well as significant correlations with important outcome variables with particular interest to military researchers. However, this study was not based on a psycholexical approach that would have required that a list of military relevant values was derived and agreed upon across the organizational hierarchy.

The research of Franke and Heinecken (2001) demonstrated that officers of different nationalities (in this particular study, within South African Military Academy cadets and US West Point cadets) differed significantly in their value orientation. With this diversity in cultural and demographic value orientations in mind, the present study examined values in the Swiss Armed Forces, which represents a military institution with a long-standing tradition to foster and cultivate values. The Swiss Armed Forces expects its members to behave in accordance with values such as responsibility, loyalty, and honesty and can therefore be seen as a value-oriented organization (Proyer, Annen, Eggimann, Schneider, & Ruch, 2012). Accordingly, the doctrine of the Swiss Armed Forces (Dienstreglement der Schweizer Armee, 2004) emphasizes the importance of personal values in successful leadership and military training (Annen, Steiger, & Zwygart, 2004). Furthermore, the recently published Swiss Federal Report on Military Ethics (Swiss Armed Forces, 2010) attributes increasing relevance to values. It is an essential identity of the Swiss Armed Forces to strengthen military values and traditions, and to foster character development. The Swiss Armed Forces, with its militia system of compulsory military service, is regarded as a mirror of society in its reflection of values. There is evidence that military experience affects political attitudes and values (French & Ernest, 1955; Jennings & Markus, 1977), having a long-lasting influence on individual-level characteristics (Jackson et al., 2012). The training goal of the Swiss Armed Forces includes distinct efforts to strengthen and influence values (Annen et al., 2004). Accordingly, these pedagogical directives make it a mandate to understand the predominant core values of the Swiss Armed Forces. More specifically, it requires a definition of those core military values that may be incorporated into the training of soldiers in the Swiss Armed Forces. Overall, core values are the stated values prioritized by an organization and help define the organization, thus giving meaning to all its members (Pathak, Rani, & Goswami, 2016). In other words, identifying organizational core values is an essential strategic instrument to inform military members and potential employees about the organization and what it represents. Within this context, Grojean and Thomas (2006) referred to the requirement to understand the predominant values within a nation's armed forces.

The Swiss Armed Forces does not have a representative definition of classified core values, as, for instance, it applies to US Army doctrine in Field Manual 22–100 with loyalty, duty, respect, selfless service, honor, integrity, and personal courage (LDRSHIP; U.S. Department of the Army, 1999, p. B-2) or to the British Army with courage, discipline, respect, integrity, loyalty, and selfless commitment (British Army Website, 2016). Accordingly, the Swiss Federal Report on Military Ethics (Swiss Armed Forces, 2010) expresses a strong interest in conducting research on values, which are in alignment with the culture of the Swiss military organization (Kernic & Annen, 2016). This current research is in line with the overall goal to develop a classification of military values as it corresponds to the Swiss Armed Forces.

7.1.4 Aims of the study

The present study addressed two prime questions, namely (1) What is the factorial structure of military values as it applies to the Swiss Armed Forces, and (2) How do the factors of the military values relate to universal values and personality? Principal component analysis in combination with Goldberg's top-down approach focused on assessing the factor structure of the corresponding military value-describing terms and on reflecting on the relationship with the Big Five personality factors. The latter was a prerequisite to interpret the outcome of the military value factors as distinct constructs. The research followed a systematic procedure aiming at developing a classification of military values in the Swiss Armed Forces, including a psycholexical analysis of military documentation, and individual ratings by military psychologists, lead commanders, Swiss career officers, and career NCOs (see the Pre-study).

7.2 Method

In this study a sample of Swiss professional officers and NCOs was tested and the structure of military values was analyzed with particular interest in the relationship between military value factors and personality traits.

7.2.1 Participants and procedure

The sample consisted of 249 career officers and 301 career NCOs of the Swiss Armed Forces ($N = 550$); 542 were males and 8 were females. Their ages ranged between 24 and 61 years ($M = 42.41$, $SD = 8.62$). Career officers included

13% Colonel, 15% Lieutenant Colonel, 11% Major, 6% Captain, and 2 First Lieutenants. Among the career NCOs were 2% Chief Warrant Officers, 8% Master Warrant Officers, 24% Staff Warrant Officers, and 22% Warrant Officers. The military professionals are assigned to various different forces or departments, with 68% working for the Land Forces, 20% for the Air Forces, and 12% for other departments, e.g., Military Security or Special Forces. This sample appears to be representative for the military professional corps. The criteria of representation were also fulfilled because we chose half of the full population to be included in the study. Take note that in Switzerland, the career officers and career NCOs are full-time military employees and are responsible for conveying military values to recruits and soldiers who, according to the conscript army system, are required to complete a certain amount of military service. About 20% of the participants indicated that they held an academic degree and almost 50% had completed vocational training as their highest educational level. The sample size clearly exceeded the size required to guarantee stability of components (cf. Guadagnoli & Velicer, 1988). For the purpose of data analysis, 39 participants with missing and unusual answers were discarded based on systematic criteria (data analysis was conducted with and without excluded data cases; only minor differences in results were observed).

The present data sampling is part of a larger research project, subject to the development of a military classification of values in the Swiss Armed Forces. Web-based questionnaires were used. The study was approved by the Chief of Swiss Armed Forces and this was communicated to the participants accordingly. A total of 783 persons received a personalized email with an invitation to participate in the online survey. Seventy-four percent of the military personnel who were initially invited completed the questionnaire. The anonymity of all participants was ensured. The military value list was the first questionnaire subsequent to the demographic assessment. Overall, the questionnaires required some 30 to 45 minutes for completion.

7.2.2 Measures

List of 25 military values as part of the MVVC

The MVVC list of 25 military values was used as derived from the psycholexical analysis in the Pre-study (e.g., esprit de corps, multiculturalism, or comradeship; see Tab. 8 for the full list). A brief list of instructions asked the participants to rate each value to the extent they are guided by that value in their everyday military decisions and actions. We explicitly advised the participants to respond in the role of their military function only, distinct from their private lives as civilians.

We used the following answer format (a 7-point scale): 1 (*I do not orient myself by this value*), 3 (*I do not orient myself often by this value*), 5 (*I do orient myself when always possible by this value*), and 7 (*I orient myself at all costs by this value*).

Universal values

Universal values were assessed by the Austrian Value Questionnaire (AVQ) by Renner, Salem and Alexandrowicz (2004). This instrument was developed on the basis of the lexical approach to account for specific facets of values in German-speaking countries. It comprises 54 items, which constitute five scales, Intellectualism, Harmony, Religiosity, Materialism, and Conservatism. Each item has to be rated on a 5-point scale as how much the person supports or disapproves it as a guiding motive in his life, ranging from 1 (*strong disapproval*), 2 (*disapproval*), 3 (*neutral*), 4 (*approval*), to 5 (*strong approval*). The first scale of Intellectualism includes cultural and humanitarian values, e.g., knowledge, individualism, or consensus. The second scale of Harmony focuses on the subjective importance of personal and social balance, e.g., sense of family, team spirit, or love. The third scale of Religiosity includes spiritual values, e.g., belief in God, forgiveness, or salvation. The fourth scale of Materialism pertains to self-centered interests, e.g., career, pride, or success. The fifth scale of Conservatism includes values that are linked to societal adjustment in a political sense, e.g., duty, sense of tradition, or defense. The AVQ has been validated in several studies and has proved to be a reliable and valid instrument for measuring universal values among the German-speaking population (Renner, 2003a; Salem & Renner, 2004). In the present sample, high internal consistency for the five scales was yielded with Cronbach alpha coefficient between $\alpha = .78$ and $\alpha = .96$.

Big Five personality traits

In addition to the list of 25 military value descriptors, participants completed the Big Five Inventory (BFI; John, Donahue, & Kentle, 1991; German Version: Lang, Lüdke, & Asendorpf, 2001). The BFI is a self-report questionnaire with a 5-point answer format for assessing the Big Five personality factors of Extraversion (8 items; e.g., "Is talkative"), Agreeableness (9 items; e.g., "Is helpful and unselfish with others"), Conscientiousness (9 items; e.g., "Makes plans and follows through with them"), Openness to experience (10 items; e.g., "Likes to reflect, play with ideas"), and Neuroticism (8 items; e.g., "Worries a lot"). It was administered to analyze the relationship between military values and personality traits. The BFI is widely used in research and has demonstrated good psychometric properties in validation studies (Rammstedt, 1997; 2007). In the present sample, the BFI yielded high internal consistency for the five scales: .80 for Extraversion, .71 for Agreeableness, .75 for Conscientiousness, .79 for Openness for Experience, and .78 for Neuroticism. As Rammstedt (1997) reported, significant correlations between the scales occurred. Extraversion was related with Openness for experience (.39, p <.01), Conscientiousness (.23, p <.01), and Agreeableness (.14, p <.01). Conscientiousness corresponded with Agreeableness (.21, p <.01) and Openness for Experience (.12, p <.01). Neuroticism related negatively with Agreeableness (-.28, p <.01), Conscientiousness (-.30, p <.01), and Extraversion (-.27, p <.01).

7.3 Results

7.3.1 Primary analyses

Descriptive statistics for the ratings of the 25 military value-describing terms are shown in Tab. 10. The means ranged from 4.46 (multiculturalism) to 6.37 (honesty), with an average mean of 5.72. Skewness and kurtosis indicated close normal distribution, except for the ratings of human dignity (Sk = -1.60 and K = 3.05) and honesty (Sk = -1.57 and K = 3.55) showing leptokurtic distributions[21]. Correlations with demographics were generally moderate in size, yet statistical significance was found between age and the military values trust, fairness, respect of the next ones, human dignity, honesty, freedom, peace, and multiculturalism. Similar relations were shown between the demographic variable years of military service and the same military values. Effects disappear when partial correlations are calculated, with age as a controlling variable. Therefore, the subsequent Pearson's r correlational analyses were controlled for a potential impact of this demographic variable.

7.3.2 Factor structure of military values

A principal component analysis (PCA) was conducted with the 25 military value descriptors. The Bartlett test of sphericity indicated that the variables fit the conditions for computing a PCA (p < .001). Factors of military values were retained based on their eigenvalues (scree test, parallel analysis of random data), on the hierarchy of factors, and according to the interpretation of the solution. The scree test did not provide a clear direction on how many factors to extract

---

21 We tested to skip these items for structural analysis with the outcome that no substantial differences in factorial structure occurred. Accordingly, items were not excluded for results reported.

Tab. 10: *Descriptive statistics of the 25 military value descriptors*

Descriptive statistics							
Military value descriptors	*M*	*Med*	*SD*	*S*	*K*	*Min*	*Max*
human dignity	6.32	7.00	0.93	-1.60	3.05	2	7
respect	6.31	6.00	0.82	-1.28	2.44	2	7
comradeship	5.75	6.00	0.99	-0.68	0.80	1	7
solidarity	5.23	5.00	1.21	-0.80	0.62	1	7
trust	6.02	6.00	1.09	-1.21	1.36	2	7
performance of duty	6.19	6.00	0.91	-1.24	2.01	2	7
performance of mission	6.33	7.00	0.85	-1.23	1.40	3	7
obedience	5.42	5.00	1.09	-0.83	1.17	1	7
integration	4.85	5.00	1.39	-0.94	0.62	1	7
hierarchy	4.89	5.00	1.38	-0.65	-0.08	1	7
security	6.17	6.00	1.02	-1.42	2.35	1	7
peace	5.43	6.00	1.34	-0.94	0.81	1	7
freedom	5.69	6.00	1.28	-1.15	1.61	1	7
justice	5.99	6.00	1.05	-1.23	2.18	1	7
obedience of the laws	5.91	6.00	1.07	-1.16	1.63	1	7
role model	6.25	6.00	0.79	-0.90	1.04	2	7
honor	5.63	6.00	1.38	-1.15	0.92	1	7
multiculturalism	4.46	5.00	1.58	-0.61	-0.25	1	7
esprit de corps	5.59	6.00	1.16	-1.01	1.32	1	7
respect of the next ones	6.02	6.00	0.97	-1.22	2.18	2	7
teamwork	5.52	6.00	0.99	-0.70	1.15	2	7
honesty	6.37	7.00	0.80	-1.57	3.55	2	7
fairness	5.79	6.00	1.05	-1.01	1.67	1	7
coherence	5.59	6.00	1.01	-0.74	1.33	1	7
autonomy	5.29	5.00	1.32	-0.80	0.54	1	7

Note. $N = 550$. M = mean, *Med* = median, SD = standard deviation, S = skewness, K = kurtosis, *Min* = minimum, *Max* = maximum.

(first eight eigenvalues were 6.81, 2.28, 1.61, 1.49, 1.22, 1.04, 0.98, and 0.81) and a parallel analysis (Horn, 1965) suggested five eigenvalues were greater than chance. These first analyses indicated that a maximum of six factors was considered to be relevant. It can be seen from the trend of the eigenvalues that there is a very potent first factor that explained 27% of variance. Keeping the findings of previous studies in mind, it was suggested to label this factor a general value factor, thus reflecting individual differences in military values preference.

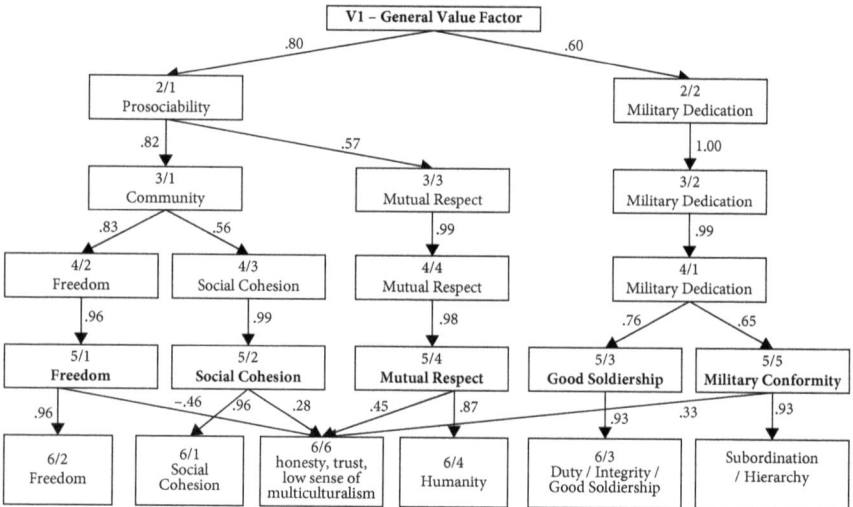

Fig. 4: The emergence of factors from the 25 military value descriptors (first unrotated principal component) starting from a general factor to six-factor solution rotated according to the varimax criterion (*N*=550); numbers within boxes indicate the number of factors extracted for a given level. Correlation coefficients to adjacent factors are only displayed when exceeding a coefficient > .20. Boldface indicates final five-factor solution

To gain further insight into the factor structure and to allow to compare several solutions, the hierarchical factor analysis procedure by Goldberg (2006) was performed. Solutions with two to six factors were extracted to show how the factors unfold, with varimax rotation in each case. The factor scores were saved for each solution and correlations between factor scores at adjacent levels were computed. The hierarchical structure from one factor (the general value factor) to six factors and the succession of factor extraction are displayed in Fig. 4 (showing correlations between the factors scores for those >.25).

The order of the factors in the solutions in Fig. 4 shows that after the general factor was split into two broader factors, one factor (2/2) emerged that was very robust throughout the fourth level of the hierarchy. Highest loadings on this broad factor were found for obedience, hierarchy, performance of duty, and performance of mission. This factor was tentatively labeled "Military Dedication." At the next lower hierarchy level this factor (4/1) splits into "Good Soldiership" (5/3), with which it has 58% of the variance in common, and "Military Conformity" (5/5), with 42% of the variance in common. The second factor (2/1) called "Benevolence and Humanity" was split into two factors at the next lower hierarchy level with which it shared 67% (3/1) and 32% (3/3) of the variance. The latter, labeled "Humanity," remains stable up to the final level of the factorial hierarchy. Factor (3/1) was split into "Benevolence" (4/2) and "Social Cohesion" (4/3) which remains identical from the four-factor solution to the five-factor solution.

The results shown in Fig. 4 indicate that the solution with more than five factors was difficult to interpret. Specifically, the new factor 6/6 from the solution of five to six factors lost substance showing only a few items (positive loadings: trust and honesty, negative loading: multiculturalism) with substantial loadings on the respective factor. Moreover, double or triple loadings on other factors were yielded. The sixth factor splits from higher level factors and is very weak, and is thus not interpretable. For this reason we decided to retain the five-factor solution. More specifically, there are five strong and one weak value factors. The five factors in the final solution explained 53.60% of the variance and could be well interpreted on the content level.

7.3.3 Five military value factors

The extracted five factors of military values are discussed in more detail as follows. Table 11 provides an overview of the military values descriptors with the highest loadings on the respective factors (five-factor solution).

Factor I: Freedom

As shown in Tab. 11, this factor covered six military value descriptors that reflect ideologies of a civic communion (peace, freedom, multiculturalism, justice, autonomy, and security). Security showed equivalent loading on factor IV (.33 [.32 on factor I]). We decided to assign security to factor I, due to the closeness in contextual meaning. Freedom seems to capture values, which reflect an ideology of a civilization and military mission. That implies a behavior, which fosters social union, peace, and security within a larger society. This factor accounted for 12.15% of the variance.

Factor II: Social Cohesion

Factor II accounted for 11.65% of the variance. The descriptors of this factor referred to group ties and social cohesion. It encompassed comradeship, coherence, solidarity, teamwork, trust, and esprit de corps. These items expressed interpersonal behavior, which typically is accepted within a larger social group, thereby demonstrating social cohesion. Only esprit de corps yielded a secondary high loading on factor III, Good Soldiership (.44 [.45 on factor II]). Esprit de corps was assigned to factor II given its explicit closeness to Social Cohesion.

Tab. 11: *Varimax loadings of the five factors based on the ratings of the 25 military value descriptors*

Military value descriptors	Factor I	Factor II	Factor III	Factor IV	Factor V	h^2
peace	**.69**	.08	-.03	.28	.25	.61
freedom	**.65**	.09	-.09	.22	.13	.51
multiculturalism	**.65**	.18	.13	.03	.07	.47
justice	**.57**	.11	.08	*.38*	.05	.49
autonomy	**.59**	.29	.12	-.07	-.03	.44
security	**.32**	.16	.19	*.33*	.28	.35
comradeship	.11	**.69**	.23	.02	.00	.54
coherence	.27	**.66**	.16	.04	.12	.55
solidarity	.27	**.63**	.08	.18	.00	.51
teamwork	.18	**.63**	.00	.12	.16	.47
trust	-.09	**.58**	.01	*.38*	.11	.50
esprit de corps	*.35*	**.45**	.44	-.15	.06	.55
performance of duty	-.01	.19	**.71**	.21	.13	.61
performance of mission	-.13	.07	**.64**	.20	.28	.55
role model	.09	.02	**.61**	.20	.13	.44
honor	.43	.19	**.55**	-.12	.08	.54
obedience of the law	.29	.03	**.37**	*.31*	.29	.40
human dignity	.14	-.04	.16	**.71**	-.07	.55
respect	.12	.22	.28	**.70**	-.04	.63
respect of the next ones	*.39*	.28	.19	**.59**	-.01	.61
honesty	-.03	*.38*	-.03	**.45**	*.34*	.47
fairness	*.31*	*.34*	-.22	**.40**	.26	.49
integration	.15	.13	.13	-.02	**.83**	.75
hierarchy	.20	.05	*.34*	-.01	**.73**	.69
obedience	.03	.14	*.51*	.00	**.64**	.69
Variance explained	12.15	11.65	10.73	10.10	8.93	

Note. N = 550. Extraction: Principal component analysis. Rotation: varimax. Boldface indicates highest factor loadings of the terms. Second loadings ≥ .30 are in *italics*. Total variance explained: 53.60 %.
Factor I = Freedom, Factor II = Social Cohesion, Factor III = Good Soldiership, Factor IV = Mutual Respect, Factor V = Military Conformity.

Factor III: Good Soldiership

This factor is summarized as Good Soldiership because it included descriptors referring to the "mindset" and characteristics of a competent military person and leader. It covered performance of duty, performance of mission, role model, honor, and obedience of the law. Honor had a high secondary loading on factor I (.43 [.55 on factor III]). Obedience of the law allowed likewise for multiple assignment, based on its similar loadings on other factors, explicitly .29 on factor I, .31 on factor IV, .29 on factor V, and a moderate .37 on factor III. Its contextual tie to Good Soldiership gave reason to assign obedience of the law to factor III. Overall, factor III accounted for 10.73% of variance.

Factor IV: Mutual Respect

Factor IV accounted for 10.10% of variance and yielded high positive loadings on human dignity, respect, respect of the next ones, honesty, and fairness. Factor I, II, and III were part of a prosocial character, with factor IV expressing personal behavior capable of successfully interacting with other people. Values with their highest loadings on this factor concerned a striving for a social equilibrium and a genuine expression of warmth and affection towards the other.

Factor V: Military Conformity

This factor accounted for 8.93% of variance and has been termed Military Conformity because the three descriptors of this factor were integration, hierarchy, and obedience. Obedience had a high secondary loading on factor III (.51 [.64 on factor V), but on ground of content-wise meaning we decided to assign it to factor V. This factor seemed to convey facets of the military regulations and hierarchy and indicated the strong special characteristics of the military organization. A high scorer on this factor could best be described as someone who likes to integrate himself in a hierarchical order of the military society.

7.3.4 Relations of the five factors of military values to universal values and personality traits

In the next step, correlations between the five factor scores of military values and the universal values of the AVQ as well as the Big Five personality traits from the BFI were analyzed. Due to intercorrelations of the military value-descriptive terms and age identified in the primary analyses, partial correlations (Pearson's r) controlling for age were computed (Tab. 12).

Table 12 shows that the majority of the correlations were numerically low to moderate ($\leq .31$). As far as the relationship to universal values was concerned,

Tab. 12: *Partial correlations (controlled for age) of the factor scores of the five-factor solution with the five universal value types (AVQ; Renner, Salem, & Alexandrowicz, 2004) and the five factors of personality (BFI; John, Donahue, & Kentle, 1991)*

	Five military value factors					
	Factor I Freedom	Factor II Social Cohesion	Factor III Good Soldiership	Factor IV Mutual Respect	Factor V Military Conformity	Multiple R
Intellectualism	.21***	.12**	.06	.24***	-.07	.39***
Balance	.15***	.23***	.07	.23***	.08	.38***
Religiosity	.08	.00	.02	.06	.09*	.17**
Materialism	.04	.04	.22***	-.03	.08*	.25***
Conservatism	.21***	.17***	.31***	.06	.22***	.47***
Multiple R	.29***	.27***	.36***	.38***	.25***	
Neuroticism	.08	-.13**	-.14**	-.05	-.12**	.24***
Extraversion	-.05	.16***	.07	.04	.02	.19**
Openness	.08*	.03	.02	.12**	-.15***	.23***
Agreeableness	.05	.19***	-.07	.25***	.10*	.39***
Conscientiousness	-.07	.09*	.28***	.19***	.18***	.39***
Multiple R	.17**	.25***	.30***	.36***	.28***	

Note. N = 550. Pearson's r correlation. * p < .05, ** p < .01, ***p < .001. Controlled for age.

the highest correlations were found between factor III Good Soldiership and Conservatism (.31) and factor IV Mutual Respect and Intellectualism (.24) as well as and Balance (.23). Moreover, findings indicated that the military factors predict all of the universal factors to a substantial extent, as indicated by the Multiple Rs in the last column of Tab. 12. Conservatism could be predicted to the highest amount by the five military values (R = .47). In turn, the universal value domains could predict factor III Good Soldiership (R = .36) and factor IV Mutual Respect (R = .38) to the highest extent and factor V Military Conformity to the lowest extent (R = .25), as indicated in the last row of Tab. 12. Moreover, correlation coefficients on the relation of military values to universal values were generally not higher than with military values relating to personality traits.

Regarding the relation to personality, the highest correlations were found between Conscientiousness and factor III Good Soldiership (.28) and between Agreeableness and factor IV Mutual Respect (.25). These results suggested that military personnel with higher levels of Conscientiousness considered their job responsibilities (e.g., performance of mission, role model, and honor) as important issues in their daily business. Unsurprisingly, findings indicated that Agreeableness positively correlated with factor IV Mutual Respect. Previous findings suggested consistently that this Big Five factor corresponded positively with Benevolence which was content-wise similar to the Mutual Respect in this current study. Accordingly, people with higher levels of Agreeableness tended to value concepts that were related to correct behavior in social interactions like human dignity, respect, and honesty.

The trait factor Extraversion correlated positively with values of Social Cohesion (factor II), including values like esprit de corps and comradeship that emphasized the aspects of community and being close-knit. Also, the factor Social Cohesion conveyed a touch of hedonism as comradeship especially was connected to hedonistic behavior, such as having a good time together. This finding was also in line with other results that have found that Extraversion is linked to Hedonism (e.g., Morales-Vives et al. 2012).

Openness correlated positively with factor IV Mutual Respect and negatively with factor V Military Conformity, although the numerical correlations were both small. A few significant negative correlations were found between Neuroticism and factor II Social Cohesion, factor III Good Soldiership, and factor V Military Conformity, but they were numerically small ($<.14$). The personality factor with the lowest correlations is Neuroticism, which was congruent with previous studies (e.g., Morales-Vives et al., 2012; De Raad & Van Oudenhoven, 2008). Factor IV Mutual Respect showed the highest significant correlations with personality traits when Multiple R was included.

Interestingly, the factor III Good Soldiership and factor V Military Conformity could be discriminated based on the correlations with the Big Five personality traits, although they split only at the last level from the four- to the five-factor solution: Factor III Good Soldiership correlated higher with Conscientiousness than factor V Military Conformity, which pointed to the fact that Good Soldiership included more facets of the conscientious and responsible mindset. Parallelly, there was a significant negative correlation between Military Conformity and Openness ($-.15$), whereas factor III Good Soldiership does not show any relation with this personality trait. In summary, although factor V split at a later level there is a distinct difference between factor V Military Conformity and factor III Good Soldiership. The factor III Good Soldiership could be recognized through its relations to the personality factor Conscientiousness. By means of correlating the factor scores for the five-factor solution with the Big Five personality traits, it was found that every factor displayed a distinct relation pattern to personality.

7.4 Discussion

To the current knowledge, this was the first study on the structure of psycholexically derived values within a military context. The aim was to assess the structure of Swiss military values and to conclude on their relationship with personality. Using Goldberg's (2006) top-down approach, the hierarchical factor structure of 25 military values was analyzed within a sample of career officers and career NCOs. Five factors of military values were identified, although there was evidence for a six-factor solution with the sixth factor showing loadings on different items. Since the interpretation of the system with six factors was less reliable, we decided to give preference to the five-factor structure. The five factors offered a robust base and could be well-interpreted; these factors were: Freedom (I) (preferring peace, freedom, justice, autonomy, and multiculturalism as guiding principles); Social Cohesion (II) (pertaining to comradeship, coherence, solidarity, teamwork, esprit de corps, and trust); Good Soldiership (III) (pertaining to performance of duty and of mission, role model, honor, and obedience of the law); Mutual Respect (IV) (being concerned to maintain human dignity, honesty, respect of others, respect, and fairness); and Military Conformity (V) (preferring integration, hierarchy, and obedience as guiding principles). It is worth mentioning that the sixth factor (positive loadings: trust and honesty, negative loading: multiculturalism) may be interpreted as the presence of a value mindset combining trust and honesty with reduced acceptance for multiculturalism and coexistence of minorities. Additionally, some value items had loadings on more than one factor. More specifically, factor I Freedom and factor IV Mutual Respect had at least three items that contributed loadings above 0.30 while being assigned to other factors. Therefore, Freedom and Mutual Respect can be qualified as general factors with the highest spread of contributive loadings across all 25 military value items. Interestingly, Freedom and Mutual Respect, which included the value items freedom, justice, human dignity, and security, reflected the prime aim and justification of the Swiss Armed Forces in providing key services to the general public. The third factor, namely Good Soldiership, was characterized by performance of duty, performance of the mission, being a role model, honor, and obedience of law, which all having reference to pro-individual principles, in contrast to the other four factors with prosocial connectivity. The remaining two factors, factor II Social Cohesion and factor V Military Conformity, are building positive emotions and focusing on structural elements of the military organization, respectively.

Relations to past findings on value factors

Without any other similar military studies, there was no validated framework to compare the outcoming structure. However, comparing this psycholexical-based military study with the military non-psycholexical study by Schumm et al. (2003), it was obvious that two factors overlap. Factor III Good Soldiership corresponds with the factor Integrity by Schumm et al. (2003) pointing to commonality in regards to being a role model, being dedicated to performance of the mission, and being duty-minded as it applies to a solid military person. Additionally, factor IV Mutual Respect aligns with Military Bearing by Schumm et al. (2003), describing a mindset of concern for others, of having high moral standards, and respect to others. On the other hand, the findings in the US Army by Schumm et al. (2003) do not correspond with the facets of Swiss military values as assessed in this sample of military professionals. The difference in the methodological approach and military culture may contribute to the difference in outcome in the two studies. More concretely, the US Army maintains its focus on operational military targets and missions, whereas the Swiss Armed Forces spends the majority of its time training for operational readiness. Accordingly, the US Army finds more dedication to loyalty and willingness to risk one's life, while social cohesion is found more in the Swiss Armed Forces (strongly reflecting the focus on military training).

Comparing the military value factors resulting from this psycholexical-based study with the universal value factors from various psycholexical studies the following can be pointed out: First, the Swiss military value factors are more specifically characterized than the universal value factors. For example, the military value factor I Freedom overlaps with the civilian value factor Broadmindness (Estonian study by Aavik & Allik, 2002), sharing the common individual values of freedom, tolerance, humanity, and multiculturalism. However, the civilian value factor of Broadmindness extends the scope, including good fortune, relaxation, and harmony as individual values. Second, it is interesting to see that distinct military factors overlap with a series of country-specific civilian value factors. For example, Factor III Good Soldiership relates to Organization and Achievement within the Dutch study. At the same time, we recognize that the Austrian value factor does not make reference to the individual values of achievement and performance of duty. Third, the military value factor IV Mutual Respect corresponds with at least one civilian value factor in each language-specific study. Correspondingly, Mutual Respect has the broadest overlap with civilian value factors. Specifically, factor IV strongly relates to Benevolence in the Spanish and Dutch studies and with Balance in the Austrian study. This finding reflected that Mutual Respect relates to a prosocial attitude in performing military service, in line with being part of a larger social union in civilian life (Wiggins, 1991). Fourth, the factor V Military Conformity does not have a strong representation in any of the systems within the other studies. However, there is an overlap with the Austrian factor Conservatism with reference to societal adjustment and discharge of duties, specific to the military context of integration, hierarchy, and obedience. The fact that the different studies were based on the same methodological approach led to the conclusion that the observation of military and civilian differences and commonalities reflect a variation in culture. More specifically, this confirmed the nature of a military-specific culture. The two most determining factors were given by the difference in sourcing the documents for the psycholexical approach (based on military documentation vs. corpus analysis from a lexicon in the respective language), and the varying size and nature of the participant sample being military or civilian. These two factors clearly influenced the corresponding value structures (Peabody & Goldberg, 1989).

Correlations of the military value factors to universal values and factors of Big Five

In this data the additional focus was on the correlation between military values and universal values measured by the AVQ. Findings showed that the universal value of Conservatism had the strongest relation to the five military value factors, especially to factor III Good Soldiership. Accordingly, people with higher levels within the five factors of military values tended to show generally a more conservative mindset, preferring value concepts which were related to duty, sense of tradition, or defense. Interestingly, correlation coefficients on the relation of military values to universal values were generally not higher than with military values relating to personality traits. This pointed to the fact that military values cannot be considered to be in close meaning to the universal values. The universal system of values turned out to be of a more comprising nature than the psycholexically derived military value factors.

An additional research question of this study concerned the correlation between military values and the Big Five personality traits to validate the distinct nature of the identified military value factors. Broadly stated, the results of this study aligned with those of previous studies (e.g., Parks & Guay, 2009; Morales-Vives et al., 2012). Specifically, the trait factors Agreeableness and Conscientiousness shared the highest amount of variance with the present military value factors, which allowed a value prediction based on personality trait. These findings corresponded to the understanding that the Big Five factors in general and specifically Agreeableness and Conscientiousness conveyed aspects of morality and could be understood as morality-based traits of personality (De Raad & Van Oudenhoven, 2008). In this study, Agreeableness correlated the highest with factor IV Mutual Respect, acknowledged as the factor which conveys most of what is interpersonally valued (Wiggins, 2003). In principle, Mutual Respect had much in common with the value factor Benevolence, which was in line with previous findings on the relationship between Benevolence and Agreeableness. Conscientiousness related to the military value factors Good Soldiership and Military Conformity, in accordance with the finding of Morales-Vives et al. (2012) who found positive relations between Conscientiousness and values of responsibility, professionalism, and duty. Accordingly, Good Soldiership and Military Conformity seemed to be similar at first glance. Consulting the correlations between the military value factors and the Big Five personality traits, factor III Good Soldiership could be recognized through its relation to the personality factor Conscientiousness and distinguished from factor V Military Conformity. However, factor I Freedom showed no substantial correlation with personality traits, marking the exception from the other value factors. Previous findings showed that Neuroticism had no substantial relation with values, which could not fully be confirmed. Neuroticism correlated negatively with factor II Social Cohesion, factor III Good Soldiership, and factor V Military Conformity, which was in line with the fact that emotional stability is an important and highly valued characteristic in the military setting and the selection process (Pols & Oak, 2007). Openness to Experience correlated positively with factor IV Mutual Respect and negatively with factor V Military Conformity. Likewise, Extraversion correlated with factor II Social Cohesion, reflecting its nature as a hedonistic value. Most of the multiple correlations between military values and personality were moderate, which suggested that values generally predict the personality traits only moderately, which was in line with previous studies.

Given the correlations between military values and Big Five traits, it can be assumed that personality has an influence to which extent the values can be accepted and adopted by the military person. At the same time, it is part of the Swiss Armed Forces principles to base training on a dedicated value culture, to convey these values, and train the core values. This implies that a military value should be adaptable by every military person, irrespective of his or her personal traits. With reference to the personality traits, Concientiousness and Agreeableness showed the strongest correlation with the five military value factors. Accordingly, the strongest embodiment of the military value culture by persons exhibiting these personality traits can be expected.

Strengths and limitations

This study initiated a novel approach to research values in the Swiss Armed Forces. Until now, this is the first military study to identify the descriptive definition of military values by means of a psycholexical approach and to determine the underlying military value factors, thereby setting the stage to better understand the military culture in the Swiss Armed Forces. One further advantage of this study is the combination of psycholexically derived and culture-sensitive military values with measures of Big Five personality traits.

Limitations are being addressed likewise. In principle, the extracted expressions from the military documentation exposed a higher degree of normative value-describing expressions, which were desirable from a military viewpoint. Accordingly, the military value-describing expressions received a high degree of acceptance by most of the military participants. Especially in this specific sample of career officers and career NCOs with a high degree of "professional socialization," subjective ratings of relevance tended to result in a right-skewed distribution. At the same time, it needs to be pointed out that using the descriptive expressions of military values captures subjective ratings of preferences primarily, rather than true behavior.

In summary, the empirical outcome of the five military value factors provides a foundation to develop a classification of military values. This points to the importance of a systematic factor analytical approach across the different organization levels, leading to a dedicated evidence-based value structure, representing the value culture in the current military organization. The challenge is for each organization to foster the best fit between personality traits, and personal values with the value culture of the organization (Annen, 2017). For the Swiss Armed Forces, this means selecting those leaders who reflect the desired value culture best and who are able to convey corresponding military training.

Study II: Assessing the Structure of Military Virtues

8 The structure of military virtues and the relation to the five factors of the VIA-IS

8.1 Introduction

Traditionally, the scientific approach in psychology has the focus on human shortcomings and problem areas. Positive psychology, by contrast, directs the emphasis on factors having a favorable influence on human life (Gable & Haidt, 2005), more precisely on positive subjective experiences, positive individual traits, and positive institutions (Seligman & Csikszentmihalyi, 2000). The underlying hypothesis stated that positive institutions allow the exposure of positive traits, namely on character strengths (Peterson, 2006). Matthews (2009) argued that the military is seen as a positive institution, qualifying as an organization that offers essential services to society, such as the education and training of young men and women in the process of becoming soldiers, and contributing to national security. Overall, the military is an institution that works for the greater good of a society, with a strong emphasis on character development, virtues, and welfare (Matthews, 2009). Accordingly, military psychology holds a long tradition in applying the aspects of positive psychology within military disciplines (Matthews, Eid, Kelly, Bailey, & Peterson, 2006b). As part of military doctrine, leadership, and training, the concepts of character strengths and virtues are regarded as extremely important (Matthews, 2008). This underlines the principal interest within military organizations to promote research on concepts of positive psychology with special focus on military virtues and their relationship to related concepts.

There is a growing amount of empirical literature addressing the relationship between character strengths in the military setting and individual adaption ability (Brooks, 2010; Cornum, Matthews, & Seligman, 2011; Matthews, Brazil, & Erwin, 2009), work and life satisfaction (Eggimann & Schneider, 2008; Proyer, Annen, Eggimann, Schneider, & Ruch, 2012), performance (Matthews, Peterson, & Kelly, 2006a), and effective leadership (Eid, Matthews, & Johnsen, 2004; O'Neil, 2007). Furthermore, a few studies examined the profile of character strengths among military samples (Banth & Singh, 2011; Cosentino & Solano, 2012; Matthews et al., 2006b). So far, the question of the structure of military-specific virtues has received less attention in research, while outside of the military domain, there has been an increasing interest in studies identifying and structuring taxonomies of universal virtues (e.g., Dahlsgaard, Peterson, & Seligman, 2005; Walker & Pitts, 2008). In particular, the question of how many virtues can be distinguished has been examined by the psycholexical approach in different cultures (e.g., Cawley, Martin, & Johnson, 2000; De Raad & Van Oudenhoven, 2011; Morales-Vives, De Raad, & Vigil-Colet, 2014) (for an overview of the systems of virtues see Tab. 5). However, there has not been a psycholexical consideration to determine the factorial structure of military-specific virtues. This study is the first to empirically assess the structure of virtues in a military organization by means of a psycholexical and factor analytic approach. Given the fact that the military organization reflects a specific culture (Meyer, 2015), it seems appropriate to evaluate how the military virtue factors correlate with factors of character strengths assessed by the VIA-IS (Peterson, Park, & Seligman, 2005). The main idea of the current study was to combine the outcome of the military virtue factors with the measure by the VIA-IS, to conduct a fine grained analysis regarding the relation between military-specific virtues and universally defined character strengths, which further leads to a deeper understanding of military culture.

8.1.1 Five factors of character strengths measured by the VIA-IS

Under the umbrella term of positive psychology, the VIA classification of strengths (Peterson & Seligman, 2004) was established for providing a manual of the positively valued traits, i.e., positively valued traits that enable the "good life" (p. 4). The VIA classification identifies twenty-four character strengths, grouped into six universal virtues, which consistently appear in philosophical and religious texts across culture and history (Dahlsgaard, Peterson, & Seligman, 2005). These universal virtues are (1) wisdom and knowledge (e.g., curiosity, love of learning); (2) courage (e.g., bravery, integrity); (3) humanity (e.g., love, kindness); (4) justice (e.g., fairness, leadership); (5) temperance (e.g., forgiveness, prudence); and (6) transcendence (e.g., gratitude, hope). Virtues are conceived as abstract concepts, whereas character strengths are seen as concrete processes and mechanisms which allow displaying the virtues in everyday life, and which can be assessed by a psychometric instrument (Peterson & Seligman, 2004). To measure these character strengths, several instruments were developed of which the *VIA-IS* (Peterson, Park, & Seligman, 2005) is the best studied and most established. The original version of the VIA-IS is in the English language. Additionally, it exists in several other languages (e.g., Croatian, German,

Japanese, and Korean). The VIA-IS is widely used in research, and demonstrates good psychometric properties.[22]

According to Peterson and Seligman (2004), their classification of the 24 character strengths under the six virtues is not a definitive one. The assignment of the strengths to the virtues was done on theoretical grounds and still needs empirical verification. They computed first exploratory factor analyses on scale level, and determined five factors, which were similar but not identical to the six virtues of the a priori classification (Peterson & Seligman, 2004). The factor named *interpersonal strengths* combined leadership, citizenship, kindness, forgiveness, fairness, and modesty. The factor of *emotional strengths* was loaded by vitality, hope, bravery, humor, love, and social intelligence. The factor representing *intellectual strengths* embraced love of learning, creativity, curiosity, and open-mindedness. The factor called *strengths of restraint* comprised prudence, persistence, self-regulation, integrity, and perspective. And the fifth factor, identified as *theological strengths*, grouped spirituality, gratitude, and appreciation of beauty. This five-factor structure could be confirmed for the German version of the VIA-IS (Ruch et al., 2010), and for the Hebrew version as well (Littman-Ovadia & Lavy, 2012).

Peterson (2006) anticipated a factor analysis based on ipsative data. Two bipolar factors emerged with the strengths being located in a full circumplex. The first factor was labeled *strengths of the heart* (e.g., spirituality, humor) vs. *strengths of the mind* (e.g., open-mindedness, persistence), and contrasted strengths entailing emotional expression vs. intellectual restraint. The second factor was named strengths *focusing on the self* (e.g., creativity, curiosity) vs. *strengths focusing on others* (e.g., fairness, leadership), and distinguished between strengths focusing on self vs. others. This two-factor solution could be reproduced for the German VIA-IS (Ruch et al., 2010), but not for the Hebrew version (Littman-Ovadia & Lavy, 2012).

Recently, McGrath (2015) suggested a three-factor virtue model. Specifically, he identified three factors, labeled *caring*, *inquisitiveness*, and *self-control*. The three-factorial structure could be confirmed across multiple measures of strengths derived from self-report data corresponding with cultural beliefs about virtue. McGrath, Greenberg, and Hall-Simmonds (2017) explored this model

---

22 Details concerning the reliability and validity of the VIA-IS were presented in Peterson and Seligman (2004); Park and Peterson (2006); Park, Peterson, and Seligman (2006); Peterson, Ruch, Beermann, Park, and Seligman (2007); Peterson, Park, and Seligman (2006); and Peterson and Seligman (2004). Details for the German version can be found in Ruch et al. (2010).

further and found substantial congruence in three-factor loadings across various samples and an overlap with measures of personality.

As Ruch and Proyer (2015) stated, factors found by running factor analysis on measures of the VIA-IS do not lead to the six virtues in the VIA classification. They pointed out that a second-order strength factor (but not a virtue) can be derived from the intercorrelations of strengths. Consequently, correlation analysis with similar concepts should be done on the level of the second-order factors of character strengths. Furthermore, the internal structure of the VIA-IS is of interest to see which character strengths co-occur within individuals and correlate with factors of military virtues.

8.1.2 Research on character strengths and virtues among military samples

Matthews (2008) introduced the concepts of positive psychology into the research domain of military psychology. A number of studies showed evidence that positive characteristics such as character strengths and virtues are critical for military leadership, predicting success, coping, and adaptation in challenging military context (Matthews, 2009). For instance, Matthews (2007) has found that those who manage to complete their education successfully at the US Military Academy scored higher on certain character strengths as compared to those who did not perform so well. Furthermore, a study by Matthews et al. (2006b) compared the VIA-IS assessed character strengths of a sample of West Point cadets with two comparison groups of Royal Norwegian Naval Academy cadets and US civilians. They found that identical character strengths (i.e., bravery, integrity, citizenship, and persistence) seemed to be important for military success in both military samples. Additionally, the West Point cadets were more similar in their rank ordering of character strengths to Norwegian cadets than they were to their own fellow American citizens. The result of this study suggested that the specific military culture plays an important role in influencing character strengths of soldiers than the difference in national origins.

In general, as Matthews (2008) outlined, the notion of Peterson and Seligman's (2004) character strengths and virtues fit well with the military's emphasis on character education, morale, values, and virtues. However, research focusing on the question of how many core virtues can be determined in the military setting is still at its beginning. There are few studies focusing on the profile of character strengths and virtues among samples of soldiers. Matthews et al. (2006b) found integrity, hope, bravery, persistence, and citizenship as the character strengths with the highest levels in two military samples from the USA and Norway. It is noticeable that three of these top character strengths (integrity, bravery, and persistence) lay within the virtue domain of courage, consolidating emotional strengths. Rather surprising, however, was the finding that kindness, humor, love, and gratitude had even higher scores than the character strength of leadership. Cosentino and Solano (2012) studied character strengths among Argentinean soldiers and showed higher levels of spirituality, social intelligence, bravery, prudence, and gratitude than in a sample of Argentinean civilians. In a study with Swiss military professionals (Eggimann & Schneider, 2008), participants scored highest on the character strengths of bravery, persistence, vitality, citizenship, self-regulation, curiosity, and hope (compared to a Swiss norm sample). This review relating military studies based on VIA-IS allowed two conclusions. The structural analysis of the military data suggested a military-specific profile of character strengths, which differed from the civilian data structure of the same nationality. This finding further confirmed the assumption that a military organization differs from a civilian institution regarding the specific military culture, and comparing the different character strengths profiles across different military organizations showed that the profile of character strength can differ due to differences in culture. It is therefore recommended to extend character strengths research to different military organizations in order to understand the military culture and the underlying core military virtues (Matthews et al., 2006b).

8.1.3 *The psycholexical approach towards the structure of virtues*

The concept of virtue has always been an essential explanatory term of scientific interest in moral philosophy (MacIntyre, 1981). Given the importance of virtues in moral disciplines, it is not surprising that definitions of virtues vary. McCullough and Snyder (2000) were interpreting a virtue broadly as "any psychological process that enables a person to think and act so as to benefit him- or herself and society" (p. 1). A more specific definition was provided by Dahlsgaard et al. (2005), defining virtues as "valued human strengths" (p. 203). This definition highlights that virtues are desirable traits, which are worth striving for (De Raad & Van Oudenhoven, 2011). In this study, a virtue was interpreted as a moral trait implying what one should be or do (De Raad & Van Oudenhoven, 2011).

In recent years, interest in study and classification of virtues has grown because of their importance in different fields of psychology (e.g., Peterson & Seligman, 2004; Sandage & Hill, 2001). Focusing on the question of how many core virtues can be distinguished, the psycholexical approach has been established as a new method of empirical examination of virtues. The lexical approach takes the lexicon in its particular language as the basis to extract positive connoted virtue-describing terms. Cawley et al. (2000) initiated the application of the psycholexical approach to study virtues and identified four factors of universal virtues in a sample of American participants. The factors were labeled *Empathy, Order, Resourcefulness,* and *Serenity*. A few years later, De Raad and Van Oudenhoven (2011) followed with their psycholexical study in Dutch. They found six factors of universal virtues, namely, *Sociability, Achievement, Respectfulness, Vigor, Altruism,* and *Prudence*. Finally, Morales-Vives et al. (2014) identified a structure of seven virtue factors in the Spanish language: *Self-confidence, Reflection, Serenity, Rectitude, Perseverance and Effort, Compassion,* and *Sociability*.

In conclusion, the structural analysis delivered four to seven universal virtue factors. Some factors of the three studies relating to different languages and cultures were comparable to each other (e.g., the Dutch and Spanish *Sociability,* and the American *Serenity*). At the same time, when comparing the descriptive content of the factors, there were substantial differences, which reflect the culture- and language-specific influences.

Supposing that descriptions of virtues can indeed differ from one culture to another (cf. Sandage & Hill, 2001), the psycholexical approach is an ideal method to reveal culture-related differences in virtues. Important is the assumption that people wish to talk about what is important to them and that the words they use for this purpose are found in the lexicon. To allow cross-cultural comparability, De Raad and Van Oudenhoven (2011) stated, "It is crucial that similar psycholexical procedures are applied in a variety of other languages/cultures" (p. 45). Given the fact that the military environment represents a distinct culture, the psycholexical approach with its principles is an appropriate method to be applied to the military setting, in order to assess the structure of military virtues, and to reach conclusions about the military core virtues. Accordingly, the current study aligned the psycholexical and factor analytic approach from other psycholexical studies with the specific culture of the Swiss Armed Forces.

It should be mentioned that military virtues were studied here in the specific environment of the Swiss Armed Forces, i.e., in a positive institution that fosters and cultivates values and virtues. The institution of the Swiss military organization expects its member to demonstrate strengths such as courage, respect, or responsibility (Proyer, Annen, Eggimann, Schneider, & Ruch, 2012). There has not yet been a factor analytic study conducted to conclude on the factorial structure and the underlying core virtues of the Swiss Armed Forces.

8.1.4 Aims of the study

The present study focused on two prime questions: (1) What is the factorial structure of military virtues as it applies to the Swiss Armed Forces, and (2) How do the factors of the military virtues relate to the second-order factors of character strengths?

The psycholexical approach was used to establish a full list of military virtues (see the Pre-study), which is then used in this study to determine the structure of military virtues in the Swiss Armed Forces. More specifically, the factor analytic approach focused on assessing the factor structure of the military virtues by administering the MVVC (see the Pre-study) to a sample of future militia officers. A further purpose of this study was to analyze how the military virtue structure corresponds with the structure of the five factors of character strengths. The latter was a prerequisite to interpret the outcome of the military virtue factors as a distinct construct, to define the Swiss core military virtues and to get a deeper understanding of the culture in the Swiss Armed Forces.

8.2 Method

In this study, a sample of Swiss officer candidates was tested and the structure of military virtues was analyzed with particular interest in the relationship between military virtue factors and factors of character strengths.

8.2.1 Participants and procedure

The sample consisted of 270 officer candidates of the Swiss Armed Forces from different training units: 263 were male and seven female. Their ages ranged between 18 and 28 years ($M = 20.59$, $SD = 1.53$). The mother tongue of the officer candidates was German, which is (compared to French and Italian) the most dominant language of the Swiss Armed Forces. Nearly half of the participants (47%) indicated that they had completed vocational training as their highest educational level, and 53% had a high school degree. Overall, 39% of the participants were currently in a relationship and a minority of 7% reported having immigrant parents. The sample size clearly exceeded the size required to guarantee stability of components (cf. Guadagnoli & Velicer, 1988). For the purpose of data analysis, 12 participants with missing values were discarded.

The present data sampling was part of a larger research project, subject to the development of a classification of military values and virtues in the Swiss Armed Forces. Web-based questionnaires were used and the anonymity of all participants was ensured. Overall, the questionnaires required 60–75 minutes for completion. The study was approved by the Chief of Swiss Armed Forces. This was communicated to the participants accordingly.

8.2.2 Measures

List of 42 military virtues as part of the MVVC

The MVVC list of 42 military virtues (e.g., loyalty, bravery, persistence) was used as derived from the psycholexical analysis in the Pre-study (see Tab. 9 for the full list). A brief instruction asked the participants to rate each virtue to the extent it applies to the self as a military person. The participants were explicitly advised to respond in the role of the military function only, as distinct from their private lives as civilians. A 7-point answer format was used with the following wording: 1 (*This virtue does not apply to me at all*), 3 (*This virtue does rather not apply to me*), 5 (*This virtue does largely apply to me*), and 7 (*This virtue does fully apply to me*). Each virtue item was presented with a short definition (e.g., "will: to want to achieve something") to prevent subjective interpretation of the term.

VIA-IS (Peterson, Park, & Seligman, 2005)

The VIA-IS consists of 240 items for the self-assessment of the 24 character strengths (10 items per strength) defined in the classification of Peterson and Seligman (2004). It uses a 5-point rating-format, ranging from 1 (*very much unlike me*), 2 (*unlike me*), 3 (*neutral*), 4 (*much like me*), to 5 (*very much like me*). A sample item is "I know that I will succeed with the goals I set for myself" (hope).

The 24 scales of the German version of the VIA-IS (Ruch et al., 2010) showed a solid reliability (median $\alpha = .77$) and high stability over nine months (median test-retest correlation $= .73$). As described in section 9.1.1, independent from the original classification of character strengths, analyses of the factor structure of the VIA-IS yielded five factors, namely, interpersonal strengths, emotional strengths, intellectual strengths, strengths of restraint, and theological strengths (e.g., Peterson & Seligman, 2004; Ruch et al., 2010). Additionally, for the German VIA-IS, Ruch et al. (2010) reported on internal consistencies ranging from .71 (honesty) to .90 (spirituality), with a median of .77.

8.3 Results

8.3.1 Primary analyses

Skewness and kurtosis of all VIA-IS scales indicated normal distribution. The means ranged from 3.08 (religiousness) to 4.00 (curiosity). High internal consistencies for the 24 scales were yielded with Cronbach alpha coefficient between $\alpha = .63$ (love) and $\alpha = .85$ (humor), and a median of $\alpha = .72$. The Bartlett test of sphericity indicated that the variables fit the conditions for computing a PCA

($p < .00$). A varimax rotated principal component analysis on scale level was computed for the VIA-IS. The five factors Ruch et al. (2010) described could be well reproduced except for fairness (assigned to strengths of restraint) and self-regulation (assigned to emotional strengths). In this sample, the five factors explained 65.28% of the variance, with the highest loadings by love and kindness on the factor of interpersonal strengths, by persistence and bravery on emotional strengths, by love of learning and open-mindedness on cognitive strengths, by modesty and prudence on strengths of restraint, and by appreciation of beauty and excellence and gratitude on theological strengths. Accordingly, the five-factor solution of the VIA-IS was included in the subsequent analyses.

Descriptive statistics for the ratings of the 42 military virtue-describing terms are shown in Tab. 13.

As shown in Tab. 13, the means ranged from 5.10 (unselfishness) to 6.25 (loyalty), with an average mean of 5.79. Skewness and kurtosis indicated close normal distribution, except for the ratings of liability ($Sk = -1.37$ and $K = 3.42$) and considerateness ($Sk = -1.32$ and $K = 2.70$) showing leptokurtic distributions. Correlations with demographics were generally low in size, yet statistical significance was found between age and the military virtues self-confidence ($r = -.14, p \approx < .05$), moderation ($r = -.12, p < .05$), and liability ($r = -.12, p < .05$). The analysis of the intercorrelation coefficients between the 42 military virtues showed that there were some distorted high intercorrelations due to duplicate terms with similar meanings. The distribution of the intercorrelations ($M = 0.21, SD = 0.13$) had a median of 0.22, with a bulge in curve between $r = .40$ and .60. Due to this higher risk of distortion of the subsequent PCA, the duplicates with intercorrelation coefficients higher than .40 were systematically analyzed. Six military virtues were skipped accordingly. Specifically, courage ($r = .68$ with bravery, $r = .55$ with boldness, and $r = .51$ with initiative and with self-confidence), willingness to perform ($r = .61$ with motivation and $r = .57$ with endurance), persistence ($r = .64$ with endurance), conscientiousness ($r = .47$ with sense of responsibility), will ($r = .58$ with motivation), and sense of duty ($r = .56$ with sense of responsibility) were excluded from the subsequent structural analysis.

8.3.2 Factor structure of military virtues

A PCA was conducted with the 36 military virtue descriptors. The Bartlett test of sphericity indicated that the variables fit the conditions for computing a PCA ($p < .00$). Factors of military virtues were retained based on their eigenvalues (scree test, parallel analysis of random data); on Velicer's minimum average partial (MAP) procedure; on the hierarchy of factors; and according to the

Tab. 13: *Descriptive statistics of 42 military virtue descriptors*

Descriptive statistics

Military virtue descriptors	M	Med	SD	S	K	Min	Max
consideration	5.43	6.00	1.01	-0.65	0.45	2	7
welfare	5.51	6.00	0.96	-0.61	0.36	2	7
faithfulness	6.14	6.00	0.80	-0.66	-0.07	4	7
personal responsibility	6.09	6.00	0.85	-0.76	0.23	3	7
liability	6.00	6.00	1.02	-1.37	3.42	1	7
discipline	6.01	6.00	1.01	-1.09	0.94	3	7
initiative	5.95	6.00	1.04	-0.97	0.89	2	7
moderation	5.35	5.00	1.15	-0.76	0.74	1	7
sense of honor	5.77	6.00	0.95	-0.53	0.13	3	7
will	6.14	6.00	0.92	-1.20	2.02	2	7
charity	5.57	6.00	1.02	-0.62	0.08	3	7
motivation	6.01	6.00	0.95	-1.12	1.16	3	7
political tolerance	5.90	6.00	1.12	-1.22	2.01	1	7
reliability	6.11	6.00	0.88	-1.06	1.73	2	7
integrity	5.70	6.00	0.95	-0.60	0.13	3	7
authenticity	6.06	6.00	0.90	-0.92	0.84	3	7
sense of duty	6.04	6.00	0.97	-1.17	1.57	2	7
conscientiousness	5.84	6.00	1.02	-1.01	1.68	1	7
credibility	6.11	6.00	0.81	-0.80	0.57	3	7
selflessness	5.52	6.00	1.14	-1.10	2.12	1	7
loyalty	6.25	6.00	0.86	-1.08	0.94	3	7
unselfishness	5.10	5.00	1.05	-0.46	0.64	1	7
bravery	5.64	6.00	1.02	-0.85	1.34	1	7
courage	5.70	6.00	1.04	-0.67	0.26	2	7
self-confidence	5.65	6.00	1.20	-1.02	1.09	1	7
modesty	5.43	6.00	1.15	-0.49	-0.28	2	7
moral courage	5.78	6.00	1.09	-0.91	0.95	2	7
boldness	5.63	6.00	1.01	-0.62	0.26	2	7
willingness to perform	5.86	6.00	1.01	-0.81	0.47	3	7
endurance	5.80	6.00	0.99	-0.86	1.14	2	7
persistence	5.95	6.00	0.98	-1.00	1.08	2	7
independence	5.81	6.00	1.01	-0.83	0.48	3	7
willingness to learn	5.80	6.00	1.05	-1.04	2.18	1	7
decision-making qualities	5.68	6.00	1.09	-1.00	1.55	1	7
moral power of judgment	5.77	6.00	0.95	-0.80	0.89	2	7

Tab. 13:   Continued

Descriptive statistics							
wisdom	5.27	5.00	1.04	-0.67	0.63	1	7
sense of responsibility	5.99	6.00	0.93	-0.94	1.21	2	7
prudence	5.76	6.00	0.92	-0.56	0.18	3	7
tenacity	5.57	6.00	1.05	-0.54	-0.13	3	7
ability to take criticism	5.63	6.00	1.13	-0.99	1.44	1	7
punctuality	6.04	6.00	1.07	-1.11	0.79	2	7
considerateness	5.72	6.00	1.09	-1.32	2.70	1	7

Note. N = 270. M = mean, Med = median, SD = standard deviation, S = skewness, K = kurtosis, Min = minimum, Max = maximum

interpretation of the solution. The scree test did not provide a clear direction on how many factors to extract. Nine eigenvalues exceeded unity. The first eigenvalues were 9.93, 2.26, 1.93, 1.51, 1.39, 1.28, 1.20, 1.14, 1.07, 0.98, 0.91, 0.83, and 0.80. In a parallel analysis (Horn, 1965), factor values were derived based on random numbers equivalent to those used in this study, showing that for the military data the first four eigenvalues were greater than chance. Velicer's (1976) MAP test determined a number of three factors to retain. These first analyses indicated that a maximum of nine factors was considered to be relevant, with scree test, parallel analysis of random data, and MAP test pointing to a solution of three to five factors.

To gain further insight into the factor structure, the hierarchical factor analysis procedure by Goldberg (2006) was performed. Solutions with two to five factors were extracted to show how the factors unfold, with varimax rotation in each case. Following the procedure described in the psycholexical study on virtues by De Raad and Van Oudenhoven (2011), the factor scores were saved for each solution and correlations between factor scores at adjacent levels were computed. The hierarchical structure from the first unrotated principal component to five factors, the construction and the stability of the factor solutions are displayed in Fig. 5 (showing correlations between the factors scores for those >.20).

As can been seen from Fig. 5, some factors at one level split into multiple factors at the next level. In a first step, the general factor was split into two broader factors, labeled "Perseverance and Courage" (2/1) and "Consistency and Empathy" (2/2). Factor 2/1 remained stable throughout the third level when it split into the robust factor "Fortitude" (4/1) with which it shared 92% of the

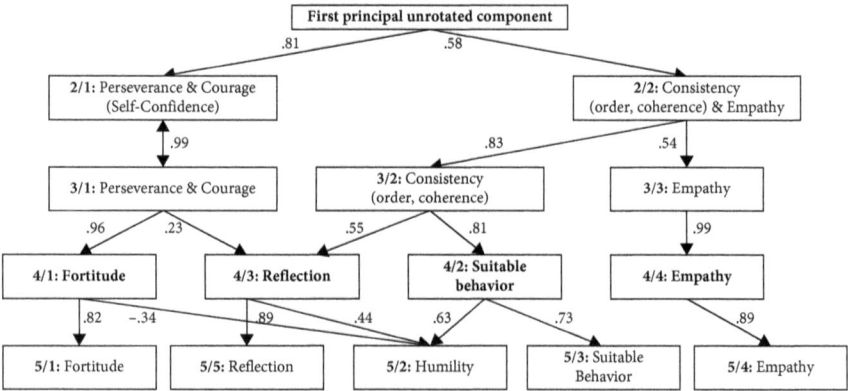

Fig. 5: The emergence of factors from the 36 military value descriptors (first unrotated principal component) starting from a general factor to five-factor solution rotated according to the varimax criterion (N=270); numbers within boxes indicate the number of factors extracted for a given level. Correlation coefficients to adjacent factors are only displayed when exceeding a coefficient >.20. Boldface indicates final four-factor solution

variance. Factor 2/2 split into two factors on the third level, with "Empathy" (3/3) remaining stable throughout the fifth level, and "Consistency" (3/2). Additionally, from factor "Consistency" (3/2) and with a partial amount of common variance from factor "Perseverance and Courage" (3/1) factor 4/3 emerged, which was labeled "Reflection." The factor "Suitable Behavior" (4/2) was constructed from factor "Consistency" (3/2), with which it shared 66% of the variance. In general, the factors at a lower hierarchical level should be seen as more specific than factors at a higher level. The results shown in Fig. 5 indicated that the solution with more than four factors was difficult to interpret. Moreover, double or triple loadings on other factors were yielded in the five-factor solution. For this reason, it was decided to retain the four-factor solution. The four factors in the final solution explained 43.45% of the variance and could be well interpreted on the content level. The factors are explained in more detail below.

8.3.3 Four military virtue factors

Table 14 provides an overview of the military virtue descriptors with the highest loadings on the respective factors (four-factor solution).

Factor I: Fortitude

As shown in Tab. 14, this factor covered military virtue descriptors that relate to positive characteristics of a strong personality expressing full motivation and zest (e.g.,

Tab. 14: *Varimax loadings of the four military virtue factors based on the ratings of 36 military virtue descriptors*

Military virtue descriptors	Factor I	Factor II	Factor III	Factor IV	h_2
(a) initiative	.66	.08	.30	.06	.54
(b) motivation	.65	.19	.10	.07	.48
(c) bravery	.65	.13	-.01	.32	.53
(d) authenticity	.61	.28	.26	.08	.52
(e) moral courage	.61	-.04	.20	.38	.55
decision-making qualities	.58	-.03	.50	-.02	.59
(f) boldness	.58	.19	-.06	.25	.44
(g) self-confidence	.57	.05	.15	.08	.35
(h) personal responsibility	.53	.41	.22	-.12	.51
endurance	.51	.26	.19	.24	.42
independence	.48	.12	.46	.00	.45
tenacity	.37	.27	.19	.17	.28
integrity	.29	.59	.33	.22	.42
moderation	-.19	.57	.24	.31	.52
liability	.34	.56	.19	-.03	.46
discipline	.39	.56	-.02	-.06	.47
considerateness	-.08	.54	.33	.31	.50
sense of responsibility	.29	.52	.40	.10	.53
faithfulness	.37	.47	-.07	.21	.40
sense of honor	.30	.41	.11	.33	.38
modesty	-.07	.41	.05	.42	.35
reliability	.33	.39	.31	.19	.40
punctuality	.22	.32	.33	-.02	.26
moral power of judgment	.17	.10	.62	.20	.46
prudence	.11	.13	.61	.05	.40
wisdom	.10	.18	.56	.21	.39
willingness to learn	.23	.05	.56	-.01	.37
ability to take criticism	-.05	.29	.43	.26	.33
credibility	.38	.35	.39	.20	.45
political tolerance	.10	.29	.38	.27	.31
consideration	-.04	.10	.15	.70	.53
welfare	.33	.05	.19	.60	.50
charity	.30	-.06	.19	.59	.48
selflessness	.37	-.03	.12	.58	.48

(continued on next page)

Tab. 14: Continued

	Factor I	Factor II	Factor III	Factor IV	
unselfishness	.06	.16	-.04	.48	.26
loyalty	.28	.27	.15	.41	.35
Variance explained	15.05	10.10	9.44	8.85	

initiative, bravery, self-confidence, moral courage). A person with high scores on this factor is able to pursue long-term goals, willing to perform on high levels, accepting responsibility, and behaving morally in difficult situations. Independence (.48 [.46 on factor III]) and decision-making qualities (.58 [.50 on factor III]) showed equivalent loadings on a second factor. This factor accounted for 15.05% of the variance.

Factor II: Suitable Behavior

Factor II accounted for 10.10% of the variance. The descriptors of this factor ensure a mode of action, which implies appropriate behavior in a military setting as reflected by the soldier showing discipline, and a sense of honor and integrity. Additionally, it encompassed liability, faithfulness, and sense of responsibility. All these items expressed an appropriate and serious behavior of a person in a leadership role maintaining order and rectitude. Some items yielded secondary high loadings on other factors, with the highest for discipline (.56 [.39 on factor I]), sense of responsibility (.52 [.40 on factor III]), and faithfulness (.47 [.37 on factor II]). Modesty showed equivalent loading on factor IV (.42 [.41 on factor II]). It was decided to assign modesty to factor II, due to the closeness in contextual meaning. Likewise, punctuality had a similar loading on factor III (.33 [.32 on factor II]) and was reassigned to factor II due to its characteristics implying an explicit appropriate behavior in the military context.

Factor III: Reflection

This factor is summarized as Reflection because it describes personal characteristics that include farsightedness and the ability to reflect and adapt. It refers to moral power of judgment, wisdom, and ability to include criticism. Credibility allowed for multiple assignment, based on its similar loadings on other factors, explicitly .38 on factor I and .35 on factor II. Its contextual tie to Reflection justified assigning credibility to factor III. Overall, factor III accounted for 9.44% of the variance.

Factor IV: Empathy

Factor IV accounted for 8.85% of variance and yielded high positive loadings on consideration, welfare, charity, selfless, unselfishness, and loyalty. It expresses

Tab. 15: *Correlations between four military virtue factors and five factors of character strengths (VIA-IS; Peterson, Park, & Seligman, 2005)*

	Four factors military virtues				
	Factor I Fortitude	Factor II Suitable Behavior	Factor III Reflection	Factor IV Empathy	Multiple R
VIA-IS I interpersonal strengths	.22***	-.03	.21**	.42***	.52***
VIA-IS II emotional strengths	.51***	.45***	.09	-.07	.69***
VIA-IS III intellectual strengths	.08	-.04	.41***	.00	.42***
VIA-IS IV strengths of restraint	-.38***	.33***	.12	.36***	.63***
VIA-IS V theological strengths	.15*	-.08	-.02	.15*	.23**
Multiple R	.69***	.57***	.48***	.58***	

Note. $N = 270$. * $p < .05$, ** $p < .01$, *** $p < .001$, two-tailed.

an interpersonal behavior, describing the ability to think of others and not be focused only on oneself. Accordingly, factor IV includes a prosocial character. Only a few secondary loadings occurred with selflessness (.58 [.37 on factor I]), welfare (.60 [.33 on factor I]), and charity (.59 [.30 on factor I]).

8.3.4 Relations of the four factors of military virtues to the five factors of character strengths

In the next step, correlations between the four factor scores for the military virtues and the five strength factors were analyzed. Table 15 shows the respective correlation coefficients between the military virtue factors and the second-order factors of the VIA-IS.

As seen in Tab. 15, the correlation coefficients ranged from $r = -.04$ to $r = .51$ with a median of .17. The highest numerical correlations were found between the emotional strengths and factor I Fortitude (.51) as well as factor II Suitable Behavior (.45). The interpersonal strengths showed a high significant correlation to factor IV Empathy of $r = .42$ and intellectual strengths to factor III Reflection (.41). The strengths of restraint were mainly linked to factor IV Empathy (.36) and to factor I Fortitude with a negative correlation of $r = -.38$. The correlation coefficients for the relationship between the theological strengths and the four factors of military virtues were very low ($r \leq .15$).

Moreover, findings indicated that the military virtue factors predict the factors of character strengths to a substantial extent ($R \geq .42$, except for the theological strengths), as indicated by the Multiple Rs in the last column of Tab. 15. The emotional strengths ($R = .69$) and the strengths of restraint ($R = .63$) could be predicted to the highest amount by the four military virtues. In turn, the strength factors could predict factor I Fortitude ($R = .69$) to the highest and factor III Reflection ($R = .48$) to the lowest extent, as indicated in the last row of Tab. 15.

Interestingly, factor I Fortitude and factor II Suitable Behavior could be discriminated based on the correlations with factors of character strengths: Although they both showed a high correlation with emotional strengths, factor I Fortitude correlated negatively with strengths of restraint and factor II Suitable Behavior showed a positive correlation coefficient. This result pointed to the fact that factor I Fortitude included facets for which strengths of restraint are not favorable. In other words, Fortitude could be characterized through its significant negative relation to strengths of restraint. Overall, results show that the military virtues and the factors of character strengths seem to overlap well.

8.4 Discussion

The present study was designed to assess the structure of Swiss military virtues and to determine how they relate to the five factors of character strengths measured by the VIA-IS (Peterson et al., 2005). Using Goldberg's (2006) top-down approach, the hierarchical factor structure of the military virtues within the MVVC was analyzed. Due to high intercorrelations between some military virtue items, six military virtues were omitted at the beginning and the main analysis was conducted with 36 military virtues. The final results were sourced from a sample of Swiss officer candidates and exposed a robust and well-interpretable solution of four military virtue factors, i.e., Fortitude (I) (with highest loadings of initiative, motivation, bravery, and authenticity); Suitable Behavior (II) (mainly pertaining to integrity, moderation, liability, and discipline); Reflection (III) (covered by moral power of judgment, prudence, wisdom, and willingness to learn as the items with the highest loadings); and Empathy (IV) (with highest positive loadings of consideration, welfare, charity, and selflessness). Some military virtue items had loadings on more than one factor. Specifically, factor I Fortitude had the most items that contributed loadings above 0.30 while being assigned to other factors. This factor accounted for the highest amount of the variance. Thus, Fortitude can be qualified as a general military virtue factor with the highest spread of contributive loadings across all military virtue items. Interestingly, Fortitude, which included the virtue items of bravery, moral courage, endurance, and decision-making qualities, reflected characteristics of a strong personality with the power to cope with challenging training and combat situations. This factor seems to be very important when considering the military environment, where decisions are often made under extreme conditions (Kornguth et al., 2010). The second and third factors, Suitable Behavior and Reflection, combine to be related to individual desirable traits of a soldier. While the factor Suitable Behavior describes the correct action of a military person, Reflection covers positive characteristics of a military person, being "mentally fit" and able to accept criticism. This factor III Reflection is considered a capability to adapt to the lessons-learned, and becoming a credible leader and behaving in line with factor II Suitable Behavior. Referring to the factor model by Peterson (2006), the first three military virtue factors represent characteristics with the focus on the "self" of a military person, while factor IV Empathy directs to the perspective of "other." Furthermore, Empathy relates to a prosocial attitude in performing military service and being part of a larger social union (Wiggins, 1991).

Relations to past findings on factors of character strengths

There were no corresponding military studies based on psycholexical defined military virtues serving as a reference to compare the findings of the four factors of military virtues. Therefore, the comparison was made with psycholexical studies conducted with universal factors of virtues (i.e., Cawley et al., 2000; De Raad & Van Oudenhoven, 2011; Morales-Vives et al., 2014). The results indicated that the military virtue factors are comparable to the factors obtained in universal virtue studies. Specifically, it was interpreted that the military virtue factors are more similar to the Dutch (De Raad & Van Oudenhoven, 2011) and Spanish universal virtues (Morales-Vives et al., 2014) than to the American virtue system (Cawley et al., 2000). For instance, there is an overlap between the military virtue factor I Fortitude, the two Dutch factors Vigor and Achievement, and the two Spanish factors Self-Confidence and Perseverance and Effort. This equivalence implies that Fortitude represents a broad scope extending to characteristics of Self-Confidence and Perseverance. Moreover, factor II Suitable Behavior corresponds with the American factor Order, the Dutch factor Respectfulness, and the Spanish factor Rectitude, which had different labels but show a strong alignment to factor II. Factor III Reflection corresponds with the Dutch factor Prudence and the Spanish factor Serenity. Factor IV Empathy shows equivalence with the American factor Empathy, the Dutch factor Altruism, and the Spanish factor Compassion. In review, all four military virtue factors are represented in the structure of the universal virtues. This indicates that the military and the universal lexical virtues show similarities in content.

In a subsequent step, the correlations between military virtues and the five factors of character strengths were analyzed. Findings demonstrated that the four military virtue factors showed the strongest relation to the emotional strengths within the VIA-IS. The strongest tie was found between both factor I Fortitude and factor II Suitable Behavior, and emotional strengths. With this correlation, factors I and II appeared to exhibit similarity. However, there was the finding that Fortitude was recognized to show a substantial negative correlation ($r = -.38$) to the strengths of restraint. In contrast, Suitable Behavior showed a positive relation to strengths of restraint. This implies, in effect, that strengths of restraint, depending on the situation, can be both beneficial and inhibitory in the military setting. It is worth mentioning that the factor III Reflection was the only military factor correlating significantly with intellectual strengths. In contrast, theological strengths had no corresponding counterparts within the structure of military virtues. Referring to the correlations between the military virtues and the character strengths, it can be concluded that they show conformity, but to a statistically moderate extent. The strengths factors from the VIA-IS are reflected in the military virtues with a specific pattern, which can be interpreted to suggest that the military virtues build on the character strengths.

Summarizing remarks

The present study has identified the structure of military virtues in a sample of Swiss officer candidates and compared it with the structure of five factors of character strengths. The results showed that the four military virtue factors are not identical to the factors of the VIA-IS, but have some similarities with the five factors of character strengths, especially with relation to the emotional strengths and strengths of restraint. Therefore, it was concluded that the military virtues, obtained with the psycholexical procedure, accommodate the universal-related factors of character strengths to some extent. However, further studies are needed to replicate the military virtue structure found in this study.

As far as limitations are addressed, most of the expressions of military virtues as part of the MVVC are highly desirable from a military viewpoint. Accordingly, these military-specific virtues tend to receive high subjective acceptance by most of the military participants; and this could result in a different answer behavior than with the non-military-specific items from the VIA-IS.

Morales-Vives et al. (2014) stated it is difficult to generalize a system of virtues from one culture to another. This was confirmed also for the relation between military virtues and the factors of character strengths, implying that there are cultural differences between the military and the civilian setting. As Morales-Vives et al. (2014) pointed out it is therefore not advisable to translate questionnaires for assessing moral traits from one culture to another. The current study may be helpful to other researchers when developing questionnaires to assess virtues in the military context, taking the factors found in this study as a starting point.

This was the first study to explore the structure of the psycholexically derived virtues within a military context. The results provided evidence for a first attempt to adapt the psycholexical and factor analytic approach to determine the military virtue factors. The culture-specific military virtues provided a good basis to define organizational core virtues factors, being in line with the specific culture of the Swiss Armed Forces. Further studies should be made, to acquire greater insight into the structure of military virtues and the differences between military and universal virtues.

Study III: Investigating the Criterion Validity of the Five Military Value Factors and the Four Military Virtue Factors

9 Can the military value factors and military virtue factors determine organizational citizenship behavior and motivation to lead?

9.1 Introduction

Predicting a person's attitude and action within a professional organization is of principle interest. Various studies have thereby taken a closer look at the role of values and virtues as morally good, positive characteristics of personality (De Raad & Van Oudenhoven, 2011; Elizur & Koslowsky, 2001), and at organizational culture (e.g., Gregory, Harris, Armenakis, & Shook, 2009). In accordance, organizational culture is defined as a set of assumptions, values, and beliefs that find shared acceptance by members of an organization (Schein, 1985). The corresponding assumption was that the underlying values translate into the behavior, since people rely on values to guide decisions and behavior (Gregory et al., 2009). In general, the role of values and virtues in work-related behaviors has received renewed interest over the past decade (Furnham, Petrides, Tsaousis, Pappas, & Garrod, 2005). Specifically, researchers have begun to further examine the effect of personal values on motivational measures such as organizational citizenship behavior, which is defined as expressing the willingness to do more than what is demanded (Ang, Van Dyne, & Begley, 2003). Fischer and Smith (2006) reinforced the importance of corresponding research, referring to the fact that employees from different sociocultural backgrounds exhibit different beliefs, value systems, and character traits within their working environment. Furthermore, Berings, De Fruyt, and Bouwen (2004) showed that values and virtues play a central role in contributing to the fit between individuals and the employment organization, based on the assumption that people will be more motivated, satisfied, and committed when their values and virtues are congruent with those emphasized in the organization. Overall, understanding the individual differences in values and virtues offers a broader capability to guide and cope with the diversity of employees (Francesco & Chen, 2004). In a value-oriented military organization such as the Swiss Armed Forces, it is particularly important to motivate and promote people who have mindsets that fit in with the values and virtues of the military organizational culture. This typically implies that they are willing to perform above and beyond the call of duty and to fulfill the organization's mission and objectives.

Values and virtues refer to stable characteristics of individuals, which are understood as positive, morally good, and desirable (De Raad & Van Oudenhoven, 2008). A value (e.g., human dignity, respect) identifies what people find important and a virtue (e.g., loyalty, modesty) is generally understood to be a morally good trait, enabling a person to live in accordance with his or her personal values (De Raad & Van Oudenhoven, 2011). In military doctrine, values and virtues are regarded as highly influential on individual adaptation ability, work and life satisfaction, performance, and effective leadership (Matthews, Eid, Kelly, Bailey, & Peterson, 2006b). This underlines the principal interest within military organizations to promote research on values and virtues as concepts of positive psychology (Peterson & Seligman, 2004). The latest Swiss Federal Report on Military Ethics (Swiss Armed Forces, 2010) attributes an increasing relevance to military values and virtues. Correspondingly, it is part of the Swiss Army's current commitment to foster military values, traditions, character development, morale, and welfare. Furthermore, within the Swiss army conscript system of an active reserve, Swiss men aged between 19 and 31 years must fulfill their military service requirement. In accordance, Swiss soldiers are citizens in uniform (Szvircsev Tresch, 2011) and they incorporate likewise the values and virtues of Swiss society, which is a unique characteristic of the Swiss military organization. The Swiss Armed Forces are thus obliged to convey the values and virtues to their military personnel as well as to the civilian society as a whole.

Within this context, an initial part of this research project developed the MVVC on a psycholexical basis, resulting in a list of 25 military value-describing and 42 military virtue-describing terms. The subsequent factorial analysis delivered a structure of five military value factors and four military virtue factors, which are reviewed in the following section.

9.1.1 Factor structure of the MVVC

The focus in Study I and Study II was to determine the factorial structure of the military values and virtues. Accordingly, the MVVC was administered to large samples of military persons in order to conduct a factorial analysis based on the individual ratings received. The outcome of the data analysis concluded on the following five factors of military values: Freedom, Social Cohesion, Good Soldiership, Mutual Respect, and Military Conformity. Table 16 summarizes the five military value factors and the corresponding values with the highest loadings derived from the principal component analysis.

Tab. 16: *Item contents with the loadings on the five factors of military values as identified by the principal component analysis*

Military Value Factor	Content
MiVa-I: Freedom	peace, freedom, multiculturalism, justice, autonomy, security
MiVa-II: Social Cohesion	comradeship, coherence, solidarity, teamwork, trust, esprit de corps
MiVa-III: Good Soldiership	performance of duty, performance of mission, role model, honor, obedience of the laws
MiVa-IV: Mutual Respect	human dignity, respect, respect of the next ones, honesty, fairness
MiVa-V: Military Conformity	integration, hierarchy, obedience

Note. All items have loadings ≥ .30 on their factor. MiVa = Miltary Value Factors (I–V).

As summarized in Tab. 16, Military Value Factor (MiVa)-I Freedom involves the individual values of peace, freedom, multiculturalism, justice, autonomy, and security, which promote a mindset and behavior that supports social unity, peace, and security. MiVa-II Social Cohesion corresponds with comradeship, coherence, solidarity, teamwork, trust, and esprit de corps, which represent interpersonal behavior that typically finds broad acceptance within a larger social community. The third factor of Good Soldiership (performance of duty, performance of mission, role model, honor, and obedience of the laws) aligns with values that describe the mindset of a competent military person. Mutual Respect as MiVa-IV comprises human dignity, respect, respect of the next ones, honesty, and fairness, which expresses a personal behavior of interacting in the correct way with other people. The fifth factor of Military Conformity encompasses values (integration, hierarchy, and obedience) that reflect conformal behavior with military guidelines, hierarchies, and organizational regulations.

Additionally, factor analytic analysis pointed to a structure of four military virtue factors: Fortitude, Suitable Behavior, Reflection, and Empathy. The corresponding virtues for the four factors are listed in Tab. 17.

As seen in Tab. 17, military virtue factor (MiVi)-I Fortitude corresponds with virtues like initiative, motivation, bravery, authenticity, and moral courage, which relate to positive characteristics of a strong personality, being able to pursue long-term goals, being willing to perform on high levels, accepting responsibility, and behaving morally in difficult situations. MiVi-II Suitable Behavior (e.g., moderation, discipline, sense of honor) ensures a mode of action that implies appropriate behavior in a military setting as reflected by showing discipline, sense of

Tab. 17: *Scale content with the loadings on the four factors of military virtues as identified by the principal component analysis*

Military Virtue Factor	Content
MiVi-I: Fortitude	initiative, motivation, bravery, authenticity, moral courage, boldness, self-confidence, endurance, tenacity, decision-making qualities, independence, personal responsibility
MiVi-II: Suitable Behavior	integrity, moderation, liability, discipline, considerateness, faithfulness, sense of honor, reliability, punctuality, political tolerance, modesty
MiVi-III: Reflection	moral power of judgment, prudence, wisdom, willingness to learn, credibility, ability to take criticism
MiVi-IV: Empathy	consideration, welfare, charity, selflessness, unselfishness, loyalty

Note. All items have loadings ≥ .30 on their factor. MiVi = Military Virtue Factors (I–IV).

honor, and integrity. The MiVi-III Reflection (e.g., wisdom, willingness to learn, modesty) describes personal characteristics that reflect farsightedness and the ability to reflect and adapt. Empathy as MiVi-IV includes the virtues of consideration, welfare, charity, selflessness, unselfishness, and loyalty. It describes the ability to think of others and not be focused only on oneself.

Due to the conceptual similarities in definition, there might be overlaps between military values and virtue factors. For instance, MiVa-III (Good Soldiership) includes content similar to MiVi-II (Suitable Behavior) with reference to "honor" (value)/"sense of honor" (virtue) and "performance of duty" (value)/"sense of responsibility" (virtue). Accordingly, the military value and virtue factors cannot be strictly separated from each other. It is important to note that these factors of military values and virtues have emerged empirically on the basis of a psycholexical and factor analytical approach, but so far have not yet been substantiated in terms of reproducibility and validity. This current study represents a first validation, investigating the criterion validity of the Swiss military values and virtues against organizational citizenship behavior (OCB) and motivation to lead (MTL).

9.1.2 OCB and MTL as validation criteria

As widely illustrated, values and virtues play an influential role in many sorts of work-related processes (Lam, Schaubroeck, & Aryee, 2002). For instance, values as guiding principles can influence how an individual perceives and interprets a given situation (Schwartz, Sagiv, & Boehnke, 2000), as well as how he or she reacts and behaves in given circumstances (Schwartz, 1999). This corresponds with the assumption that values have a direct effect on behavior (see Kluckhohn, 1951; Rokeach, 1979b). Accordingly, the main scope of this particular study was to examine the effects of the identified military values and virtues on OCB and on MTL as it applies to recruits undergoing basic military training in the Swiss Armed Forces.

The identification of capable militia cadre among Swiss soldiers hinges on OCB and MTL. OCB and MTL have been confirmed to be crucial for selection of Swiss military officers (Annen, Goldammer, & Szvircsev Tresch, 2015). It is therefore equally essential to analyze the criterion validity of the military values and virtue factors with regards to OCB and MTL, representing two validation criteria.

Organizational citizenship behavior

The following example illustrates the concept of OCB in the military context: Marcel is a soldier in the infantry recruit school. During the perseverance-practice, he helps his comrade carry the backpack, which allows his comrade to successfully cross the finish line. With this behavior, Marcel demonstrates OCB. Citizenship behavior is often performed by employees to support the interests of the group or organization even when the behavior may not directly lead to individual benefits. This example illustrates the fact that the assistance is helpful to the military organization, although this type of help is not part of a formal requirement (Moorman & Blakely, 1995).

In recent years, OCB has gained increasing attention from organizational psychologists (Borman & Penner, 2001). The meta-analysis of Podsakoff, Whiting, Podsakoff, and Blume (2009) confirmed that OCB represents a motivational behavior pattern generating positive effects on organizational (e.g., productivity, efficiency, reduced costs, and customer satisfaction) and individual results (e.g., better managerial ratings of employee performance and higher reward allocations). Accordingly, OCB has been one of the most studied concepts in recent decades (Lepine, Erez, & Johnson, 2002; Podsakoff, Mackenzie, Paine, & Bachrach, 2000). In the initial definition by Organ (1988), OCB is described as an individual discretionary behavior that goes beyond the job description and that is for the benefit of the organization. Organ (1988) specified that OCB is not an enforceable job requirement, thus it is not directly or explicitly recognized by the formal reward system. However, because of the conceptual difficulties in describing organizational behavior, especially when it comes to the aspects of "discretionary" and "non-contractual rewards," Organ (1997) redefined OCB as "contributions to the maintenance and enhancement of the social and psychological context that supports task performance" (p. 91).

In sum, OCB can be described as the willingness to do more than what is normally demanded. It is a voluntary behavior in the workplace, which has a positive effect on the functioning capacity of the organization. OCB has become an important research topic within the military context, where it is particularly challenging to attract capable volunteers for military cadre positions. Annen et al. (2015) examined the effects of OCB on cadre career decisions in the Swiss Armed Forces, and the extent that OCB can be predictive for a military cadre career. Results showed that OCB related to both the selection as cadre and the voluntariness of a recruit to pursue a career as militia cadre, pointing to the importance of this concept in predicting significant additional effort in the military context.

If we take a closer look at the content of OCB, it involves actions such as being helpful and cooperative, tolerating inconveniences at work, taking on additional responsibilities, and keeping up with company affairs (Organ, Podsakoff, & Mackenzie, 2006). Organ (1988) proposed the five underlying dimensions to the OCB concept:

1. altruism, e.g., helping other employees with work problems,
2. conscientiousness, e.g., dealing particularly accurately with job tasks and duties,
3. courtesy, e.g., informing or consulting others before actions are taken that might affect them,
4. sportsmanship, e.g., tolerating or accepting organizational inconveniences without complaining about them, and
5. civic virtue, e.g., taking an active part within the organization and being involved in the organizational processes.

Organ et al. (2006) referred to the conceptualizations of OCB directly on theory and research regarding prosocial organizational behavior (i.e., behavior that benefits others, rather than oneself; O'Reilly & Chatman, 1986). Other similar concepts have been developed to describe this kind of organizational behavior pattern, such as organizational spontaneity (George & Brief, 1992), contextual performance (Borman & Motowidlo, 1993), or extra-role behavior (Van Dyne & LePine, 1998). Accordingly, it is obvious that OCB can be associated with a behavior based on particular values and virtues. In fact, prior research on antecedents of OCB has argued that OCB is motivated by prosocial values such as care or loyalty held by particular individuals (Halbesleben, Bolino, Bowler,

& Turnley, 2010; Organ, 1988; Organ et al., 2006;). Consistent with this behavioral notion, studies have found that employees who are less individualistic and who show concern for others have a greater tendency to engage in OCB behavior (McNeely & Meglino, 1994; Moorman & Blakely, 1995). Other studies have also provided evidence that values exert a direct effect on the behavior of individuals at work (Liu & Cohen, 2010). Prosocial values motives were most strongly linked to OCB directed at individuals, and organizational concern motives were strongly associated with OCB directed towards the organization (Rioux & Penner, 2001). Therefore, an essential motive to demonstrate OCB is the concern about the well-being of other people and a desire to be helpful and cooperative. Consequently, Clary et al. (1998) stated that both prosocial and OCB motives may be called "value-expressive" motives. Accordingly, the person is motivated by the values that he or she holds, with the underlying premise that people often choose to engage in OCB because it meets certain personal needs (Liu & Cohen, 2010).

Motivation to lead

In addition to showing OCB, Marcel is willing to voluntarily pursue a career as a militia cadre and to proceed with officer training courses after the military basic training. He is therefore motivated to take a cadre position, demonstrating MTL, and hopes to be selected for the process of becoming a future officer of the Swiss Armed Forces. MTL is particularly important within the Swiss Armed Forces, given its system of military conscripts. As this example illustrates, MTL in the Swiss Armed Forces is defined as the voluntary pursuit of a militia cadre career, which is an essential prerequisite for recruits to be eligible for selection for officer candidate training (Annen et al., 2015).

Overall, a number of studies have suggested that universal values influence motivation in the form of organizational behavior (Ajzen, 1991), goal setting and task performance (Locke & Latham, 1990), and work motivation (Vroom, 1964). In reference to the military context, Thomas, Dickson, and Bliese (2001) demonstrated that the behavior sets of effective leaders are determined by personal values and motives. Studies have shown that values such as power, affiliation, and motivation to lead are correlated with military leader performance (Van Iddekinge, Ferris, & Hefner, 2009; Thomas et al., 2001). This effect was found in addition to personality characteristics such as Extraversion and Conscientiousness. Chan and Drasgow (2001) developed an individual differences construct defined as MTL, which describes a person's efforts to assume leadership training, roles, and responsibilities. The military study by Clemmons and Fields (2011) analyzed the role of universal values as determinants of MTL and showed that values, mainly self-enhancement values, made significant incremental contributions in explaining MTL. In summary, studies highlighted that personal values, along with personality, play a distinct role in predicting MTL. As mentioned previously, the role of military values and virtues to determine MTL has not yet been examined in studies conducted in the specific context of a military organization.

9.1.3 Aims of the study

The present study had the goal to answer two prime questions: (1) To what extent can universal values, military values, and military virtues determine OCB and the motivation to pursue a MTL, and (2) What is the incremental validity of military values and virtues above universal values to determine OCB and MTL? In addition, this investigation represented a validation of the identified military value and virtue factors within Study I and II.

This study used the lexical-derived inventories of universal values, military values, and military virtues, specifically the five factors of universal values (Intellectualism, Harmony, Religiosity, Materialism, and Conservatism) by Renner (2003a); the five military value factors (Freedom, Social Cohesion, Good Soldiership, Mutual Respect, and Military Conformity); and the four military virtue factors (Fortitude, Suitable Behavior, Reflection, and Empathy). According to evidence from previous studies, the hypothesis suggested that universal, military values and military virtues predict OCB and MTL to a substantial extent. Additionally, it was assumed that the psycholexically derived military values and military virtues improve the determination of OCB and the motivation to pursue a career as militia cadre beyond the contribution of universal values.

9.2 Method

9.2.1 Participants and procedure

The present sample consisted of 396 recruits in the fifth week of their basic Swiss military training. The mother tongue of the recruits was German, which is the official language of the Swiss Armed Forces. The branch of service was infantry and participants were exclusively male (100%). Their ages ranged between 18 and 25 years ($M = 20.12$, $SD = 1.13$). Most of the participants (77%) indicated that they had completed vocational training as their highest educational level and 12% had a high school degree. Overall, nearly half of the participants (47%) were currently in a relationship and a minority of 14% reported having immigrant parents. For the purpose of data analysis, 20 participants with missing information were excluded. Data analysis was conducted with and without excluded data cases. Only minor differences in results were observed (reported data refers to the results with excluded cases).

This study was conducted as part of a larger project addressing motivation and cadre selection in the Swiss Armed Forces. It was approved by the Chief of Armed Forces. All participants gave informed consent to answer the questionnaires and were surveyed at a point of the training course prior to knowing whether they would be selected for a cadre position. Web-based questionnaires were used and the anonymity of all participants was ensured.

9.2.2 Measures

Independent variables

Universal values were assessed by the Austrian Value Questionnaire (AVQ) by Renner, Salem, and Alexandrowicz (2004). This instrument was developed on the basis of the lexical approach to account for specific facets of values in German-speaking countries. It comprises 54 items that constitute five scales, Intellectualism, Harmony, Religiosity, Materialism, and Conservatism. Each item was rated on a 5-point scale, indicating how much the person supports or disapproves it as a guiding motive in his life, ranging from 1 (*strong disapproval*), 2 (*disapproval*), 3 (*neutral*), 4 (*approval*), to 5 (*strong approval*). The first scale of Intellectualism includes cultural and humanitarian values, e.g., knowledge, individualism, or consensus. The second scale of Harmony focuses on the subjective importance of personal and social balance, e.g., sense of family, team spirit, or love. The third scale of Religiosity includes spiritual values, e.g., belief in God, forgiveness, or salvation. The fourth scale of Materialism pertains to self-centered interests, e.g., career, pride, or success. The fifth scale of Conservatism includes values that are linked to societal adjustment in a political sense, e.g., duty, sense of tradition, or defense. The AVQ has been validated in several studies and has proved to be a reliable and valid instrument for measuring universal values among the German-speaking population (Renner, 2003b; Salem & Renner, 2004). In the present sample, high internal consistency for the five scales was yielded with Cronbach alpha coefficient between $\alpha = .85$ and $\alpha = .96$.

To assess military values and virtues we used the MVVC, which was derived from a psycholexical analysis of military documentation and developed in a systematic stepwise procedure (see the Pre-study). It was established to assess the specific facets of values and virtues in the Swiss military culture. The inventory consists of two sections. Composite scale measures for both sections were computed by summing across items within each factor, with higher scores indicating greater endorsement of the respective value factor. Additionally, mean values within each factor were computed to use for descriptive statistics. Since this is a newly developed catalog, further studies have not yet been conducted in applying this measure and assessing military values and virtues in other military samples.

Section 1 of the MVVC comprises 25 military values (e.g., esprit de corps, multiculturalism, or comradeship), and instructs the participant to rate each item to the extent they are guided by that value in their everyday military decisions and actions. Each item is presented with a short definition (e.g., performance of duty: fulfill your tasks reliably) to control for subjective interpretation of the term. A 7-point scale running from 1 (*I do not orient myself by this value*), 3 (*I do not orient myself often by this value*), 5 (*I do orient myself when always possible by this value*), to 7 (*I orient myself at all costs by this value*) was used.

The participants are explicitly advised to respond in the role of their military function only, distinct from their private lives as civilians. Five factors can be constituted: Freedom, Social Cohesion, Good Soldiership, Mutual Respect, and Hierarchy. In the present sample, internal consistency for the five computed scales of military values was yielded with Cronbach alpha coefficient between α =.79 and α =.85.

Section 2 of the MVVC comprises 42 military virtues (e.g., loyalty, integrity, or modesty), with the instruction to rate each virtue to the extent it applies to the self as a military person. Each item is presented with a short definition (e.g., welfare: efforts for the benefit of poor and needy people) to prevent subjective interpretation of the term. The 7-point scale ran from 1 (*This virtue does not apply to me at all*), 3 (*This virtue does rather not apply to me*), 5 (*This virtue does largely apply to me*), to 7 (*This virtue does fully apply to me*). Factor analytic analysis pointed to a structure of four military virtue factors: Fortitude, Suitable Behavior, Reflection, and Empathy. Cronbach alpha coefficients for the four computed scales of military virtues ranged between α =.72 and α =.90. Due to high intercorrelations, six expressions out of the 42 military virtues with coefficients higher than .40 were excluded (i.e., courage, willingness to perform, persistence, conscientiousness, will, and sense of duty). The subsequent analysis as part of this chapter was based on the 36 military virtues.

Criterion variables

We used a military adapted 5-item version to measure OCB, based on the OCB scale described by Meierhans, Rietmann, and Jonas (2008): "Evaluate the following statements…" (a) "I follow military rules and instruction, even when nobody is watching me," (b) "I inform myself about the weekly program," (c) "I help a comrade if he doesn't manage to keep up during an exercise," (d) "I participate actively during group assignments," and (e) "I have a high level of self-discipline and fulfill my duties properly even when I'm not being monitored/inspected." Recruits rated their agreement to the statements on a 4-point Likert scale ranging from 1 (*I completely disagree*), 2 (*I tend to disagree*), 3 (*I tend to agree*), to 4 (*completely agree*). A mean OCB scale was computed for all cases with ratings on the five items. Internal consistency in the present sample was satisfying with α = .77.

MTL was measured by the four items taken from the military recruiting test battery of the Swiss Armed Forces. "Evaluate the following statements ..." (a) "I am basically motivated to lead," (b) "I feel confident of taking a leadership position in the Swiss Armed Forces," (c) "I can imagine holding a cadre position in the near future," and (d) "I will do everything possible to get in a higher leadership position in the Swiss Armed Forces." Items were presented in Likert format using a 4-point response scale that ranged from 1 (*I completely disagree*), 2 (*I tend to disagree*), 3 (*I tend to agree*), to 4 (*I completely agree*). The Cronbach alpha coefficient for the scale was α =.92.

9.3 Results

Prior to the main analyses, multicollinearity of the independent variables was inspected. Multicollinearity of the variables could be excluded since values of variance inflation factors (VIF) and tolerance were close to 1, except for MiVi-II Suitable Behavior with a VIF of 4.26. MiVi-II was tested by being omitted for regression analysis with the outcome that no substantial differences in criterion validity occurred. Accordingly, this factor was not excluded for results reported.

Subsequently, a multiple regression analysis was conducted, regressing (a) the five universal value factors, (b) the five military value factors, and (c) the four military virtue factors on OCB and MTL as two separate criterion variables. The incremental validity of military values and virtues above universal values to determine OCB and MTL was examined with hierarchical regression analysis. The five universal value factors were regressed in a first step on OCB and MTL, followed by the block of five military values and four military virtues in a second step.

9.3.1 Primary analyses

The means, standard deviations, and correlations among the study variables are shown in Tab. 18.

Tab. 18: *Means, standard deviations, and correlations among the study variables in the sample of Swiss recruits*

Variables	Mean[1]	SD	1	2	3	4	5	6	7	8	9	10	11	12	13	14	15	16
1. OCB	2.87	0.50	-															
2. MTL	1.98	0.95	.41	-														
3. Intellectualism (UniVa-I)	3.69	0.59	.36	.16	-													
4. Harmony (UniVa-II)	4.17	0.58	.25	-.05	.69	-												
5. Religiosity (UniVa-III)	2.94	1.12	.08	.10	.23	.18	-											
6. Materialism (UniVa-IV)	3.97	0.60	.23	.06	.61	.74	.11	-										
7. Conservatism (UniVa-V)	3.61	0.77	.34	.28	.47	.44	.31	.47	-									
8. Freedom (MiVa-I)	3.15	1.01	.33	.14	.42	.35	.08	.31	.17	-								
9. Social Cohesion (MiVa-II)	3.27	0.96	.49	.23	.37	.41	.10	.33	.34	.56	-							
10. Good Soldiership (MiVa-III)	3.01	1.04	.61	.40	.36	.31	.13	.31	.49	.51	.65	-						
11. Mutual Respect (MiVa-IV)	3.30	0.96	.45	.15	.45	.45	.05	.35	.26	.67	.71	.61	-					
12. Military Conformity (MiVa-V)	2.82	1.13	.54	.29	.27	.25	.13	.23	.34	.47	.57	.73	.49	-				
13. Fortitude (MiVi-I)	2.98	0.97	.50	.37	.46	.37	.08	.38	.39	.58	.57	.64	.56	.53	-			
14. Suitable Behavior (MiVi-II)	2.79	0.97	.55	.33	.43	.37	.08	.35	.41	.62	.64	.73	.64	.68	.81	-		
15. Reflection (MiVi-III)	2.94	0.95	.45	.28	.45	.35	.04	.37	.29	.60	.52	.56	.60	.48	.84	.77	-	
16. Empathy (MiVi-IV)	2.93	0.97	0.52	.24	.47	.42	.13	.30	.31	.57	.62	.56	.64	.55	.67	.73	.68	-

Note. $N = 396$. Correlations larger than .09 are significant at $p < .05$, correlations larger than .12 are significant at $p < .01$; UniVa = Universal Values (I–V); MiVa = Military Values (I–V); MiVi = Military Virtues (I–IV). [1] In order to provide comparability of the measures, means of the military value and virtue factors were transformed from a 7-point scale to a 4-point scale.

As can be seen in Tab. 18, the intercorrelations among the five factors of universal values (mean $r = .43$, $p < .01$), the five factors of military values (mean $r = .58$, $p < .01$), and the four factors of military virtues (mean $r = .75$, $p < .01$) were positive and significant. Military value factors and military virtue factors (mean $r = .59$, $p < .01$) showed a stronger relation to each other than universal values to military values (mean $r = .29$, $p < .01$), and to military virtues (mean $r = .32$), respectively. The two measures for OCB and MTL correlated with $r = .41$ ($p < .01$). The highest correlations were found for OCB with MiVa-III Good Soldiership ($r = .61$), MiVi-II Suitable Behavior ($r = .55$), and MiVa-V Military Conformity ($r = .54$). Generally, OCB showed higher correlation coefficients with universal value factors, military value factors, and military virtue factors (mean $r = .41$) as compared to MTL (mean $r = .21$). Moreover, the results showed preliminary indications that the five military value factors and four military virtue factors were positively related to OCB and MTL to a higher extent than the corresponding results for universal value factors. Several small correlations between the demographic and study variables appeared within the sample; for example, age went along with higher levels of OCB and MTL, and higher preference of MiVa-I Freedom and MiVi-IV Empathy.

9.3.2 Universal values, military values, and military virtues as determinants of OCB and MTL

To test whether universal values, military values, and military virtues might have an effect on OCB and MTL, a multiple regression analysis was computed. The results of the multiple regression analyses, regressing the five universal values, the five military values, and the four military virtues on OCB and MTL, are reported in Tab. 19.

As displayed in Tab. 19, the R^2 showed that universal values could explain 17% of the variance in OCB and 15% of the variance in MTL. The five military values explained 41% of the variance in OCB and 17% in MTL. Similar results were shown for the four military virtues, predicting 35% of the variance in OCB and 16% in MTL.

Looking at the standardized (std) β coefficients of the five universal values, results showed that UniVa-I Intellectualism ($\beta = .29$, $p < .001$ for OCB; $\beta = .24$, $p < .001$ for MTL) and UniVa-V Conservatism ($\beta = .26$, $p < .001$ for OCB; $\beta = .31$, $p < .001$ for MTL) contributed to the explanation of both criteria, OCB and MTL. UniVa-II Harmony showed a negative std β coefficient ($\beta = -.39$, $p < .001$) to determine MTL. Consequently, Hypothesis 1 was confirmed. Among the five military value factors, MiVa-III Good Soldiership exhibited the highest criterion validity for OCB ($\beta = .38$, $p < .001$) as well as MTL ($\beta = .47$, $p < .001$).

Tab. 19: *Summary of multiple regression analysis for universal values (AVQ; Renner, Salem, & Alexandrowicz, 2004), military values, and military virtues determining organizational citizenship behavior (OCB; Meierhans, Rietmann, & Jonas, 2008) and motivation to lead (MTL; Swiss Armed Forces, 2012)*

Independent variables	OCB			MTL		
	B	SE B	β	B	SE B	β
Universal Values						
UniVa-I: Intellectualism	.25	.06	.29***	.39	.11	.24***
UniVa-II: Harmony	.00	.07	.03	-.63	.13	-.39***
UniVa-III: Religiosity	-.03	.02	-.06	.01	.04	.01
UniVa-IV: Materialism	-.06	.06	-.07	.08	.12	.05
UniVa-V: Conservatism	.17	.04	.26***	.38	.07	.31***
R^2		.17			.15	
F		16.14***			13.46****	
Military Values						
MiVa-I: Freedom	-.01	.00	-.09	.00	.01	-.02
MiVa-II: Social Cohesion	.01	.01	.11	.01	.01	.03
MiVa-III: Good Soldiership	.04	.01	.38***	.08	.01	.47***
MiVa-IV: Mutual Respect	.01	.01	.12	-.03	.02	-.15
MiVa-V: Military Conformity	.03	.01	.18**	.00	.02	.01
R^2		.41			.17	
F		53.30***			15.98***	
Military Virtues						
MiVi-I: Fortitude	.01	.00	.18	.03	.01	.44****
MiVi-II: Suitable Behavior	.02	.00	.34***	.01	.01	.14
MiVi-III: Reflection	-.01	.01	-.12	-.02	.01	-.16
MiVi-IV: Empathy	.02	.01	.23***	-.01	.01	-.05
R^2		.35			.16	
F		51.88***			18.77***	

Note. N = 396. OCB = organizational citizenship behavior. MTL = motivation to lead. ** $p < .01$, *** $p < .001$.

MiVa-V Military Conformity yielded significant std β coefficient (β = .18, $p < .01$) to contribute to the determination of OCB. With regards to the four military virtue factors, MiVi-II Suitable Behavior (β = .34, $p < .01$) and MiVi-IV Empathy (β = .23, $p < .01$) were the only two factors to determine OCB. MTL was accounted by only for MiVi-I Fortitude (β = .44, $p < .01$). Generally, military value and virtue factors provided better results of criterion validity with reference to OCB than to MTL.

Tab. 20: *Summary of hierarchical regression analysis for universal values, military values, and military virtues determining organizational citizenship behavior (OCB; Meierhans, Rietmann, & Jonas, 2008) and motivation to lead (MTL; Swiss Armed Forces, 2012)*

Independent variables	OCB			MTL		
	B	R^2	ΔR^2	β	R^2	ΔR^2
Step 1[a]		.17***			.15***	
UniVa-I: Intellectualism	.29***	.11		.24***		
UniVa-II: Harmony	.00			-.39***		
UniVa-III: Religiosity	-.06			.01		
UniVa-IV: Materialism	-.07			.05		
UniVa-V: Conservatism	.26***			.31***		
Step 2[b]		.46***	.29***		.29***	.14***
MiVa-I: Freedom	-.17**			-.10		
MiVa-II: Social Cohesion	.06			.02		
MiVa-III: Good Soldiership	.32**			.28***		
MiVa-IV: Mutual Respect	.06			-.07		
MiVa-V: Military Conformity	.15*			.01		
MiVi-I: Fortitude	.12			.36***		
MiVi-II: Suitable Behavior	.01			-.04		
MiVi-III: Reflection	-.03			-.07		
MiVi-IV: Empathy	.17**			.04		

9.3.3 Incremental validity of military values and military virtues for determining OCB and MTL

Since universal values as well as military values and virtues resulted in showing criterion validity with regards to OCB and MTL, in the next step, we tested the incremental validity of military value and virtues among and above universal values. The results of the hierarchical regression analyses determining OCB and MTL with the five universal values in a first step, and the five military values and four military virtues in a second step, are presented in Tab. 20.

As Tab. 20 indicates, inspection of the std β coefficients obtained for the first step demonstrated the same results as in Tab. 19, namely that OCB and MTL were significantly determined by UniVa-I Intellectualism and UniVa-V Conservatism, whereas UniVa-II Harmony was a negative predictor for MTL. In the first step, 17% of variance in OCB and 15% in MTL was determined. However, criterion validity was substantially improved by adding military values and virtues in a second step, increasing the total explained variance to 46% for OCB and 29% for MTL. MiVa-III Good Soldiership ($\beta = .32$, $p < .01$), MiVa-V Military Conformity ($\beta = .15$, $p < .05$), and MiVi-IV Empathy ($\beta = .17$, $p < .01$) predicted OCB positively, and MiVa-I Freedom negatively ($\beta = -.17$, $p < .01$); whereas valuing MiVa-III Good Soldiership ($\beta = .28$, $p < .001$) and MiVi-I Fortitude ($\beta = .36$, $p < .001$) contributed positively to determine MTL.

9.4 Discussion

The present study set out to clarify the effects of universal values, military values, and military virtues on OCB and MTL within the Swiss Armed Forces (representing a conscription system). Furthermore, we examined the incremental contribution of military value and military virtue factors beyond universal values to determine OCB and MTL. It is worth mentioning that this was the first study to validate the criterion validity of the psycholexically and factor analytically based military value and military virtue factors on OCB and MTL. It specifically concerned the five military value factors Freedom, Social Cohesion, Good Soldiership, Mutual Respect, and Military Conformity and the four military virtue factors Fortitude, Social Behavior, Reflection, and Empathy as assessed by the MVVC. As part of the multivariate analyses referring to a sample of recruits in the fifth week of military basic training, we looked at the two criterion variables of OCB and of the motivation to voluntarily pursue a career as a militia cadre (MTL). The following section summarizes the main results:

First, the universal values, and the five military value factors and four military virtue factors have proven to be significant variables in reference to determine OCB and MTL. However, the five military value factors and four military virtue factors showed a stronger criterion validity on OCB and MTL than universal value factors in reference to the multiple regression analysis. Furthermore, military value and virtue factors provided better results of incremental validity with reference to OCB than to MTL.

Second, military values and virtues could explain incremental variance in OCB and MTL, beyond the effects of the universal values. Specifically, criterion validity was substantially improved by adding military values and virtues beyond the universal values, increasing 29% of the explained variance in favor of OCB and 14% of MTL.

Third, the criterion validity of OCB and MTL was particularly determined by specific universal and military values and virtues. Among universal values, Intellectualism and Conservatism could be identified as positive predictors for OCB and MTL. Harmony showed a negative contribution in determining MTL.

Good Soldiership turned out to be a strong determinant of OCB and MTL. Among the military virtues, Social Behavior showed significant criterion validity for OCB, as did Fortitude for MTL.

Previous studies showed a direct impact of values and virtues on employees expressing OCB and MTL (e.g., Halbesleben et al., 2010). Our study confirmed these findings, verifying that universal values as well as military-specific values and virtues play an essential role in determining work-related behavior such as OCB and MTL. The findings allow researchers to conclude that military values and virtues showed a higher contribution to determine OCB than universal values. We interpret this additional criterion validity of military values and virtues by the culture-sensitivity of the military factors. Recent research dealing with an individuals' fit and coherence within the organization referred to the beneficial effect on employees whose values match with those of the organization or with those of some specific individuals within the organization (Chatman, 1991; Meglino & Ravlin, 1989; Meglino, Ravlin, & Adkins, 1991).

However, another reason for the high validity of military values and virtues may be given by the context-specific characteristics of the identified military value and virtue factors, producing a common variance between the measures (Breckler, 1990). Strack, Gennerich, and Hopf (2008) pointed to the fact that the question of context-specific assessment of values and virtues is not adequately solved yet. Consequently, it could not be answered conclusively if the culture-sensitivity or the context-specificity of the measures was driving the effects found in this study.

Although military values and virtues could significantly predict both OCB and MTL criteria, OCB could be determined to a higher extent than MTL by military values and virtues. This supports the evidence that OCB reflects a value- and virtue-expressive behavior, in line with Clary et al. (1998), stating that motives to show OCB may be "value-expressive" motives. An important criterion is the concern for the welfare of other people and a desire to be helpful and cooperative. If we look at the conceptual notion of OCB, a good citizen is an employee who offers support to the organization even when no such support is or can be expressly required (Organ et al., 2006). According to O'Reilly and Chatman (1986), similar concepts to OCB are prosocial behavior patterns of which the primary benefit is with others, rather than themselves. Accordingly, OCB can be associated with a prosocial behavior based on certain values and virtues, as demonstrated by these data. Interestingly, Harmony was not related to OCB although this factor includes aspects of personal and social balance, e.g., team spirit or love. However, the universal value of Intellectualism is covering humanitarian values such as consensus and consideration, which explains the contribution explaining variance in OCB and MTL. The result that Conservatism showed criterion validity towards OCB and MTL points to the fact that a recruit's conservative mindset promotes behaviors of additional efforts as measured by OCB and the motivation to pursue a career as militia cadre. Additionally, Good Soldiership was confirmed to be a strong predictor for OCB as well as for MTL. This particular military value factor reflects the mindset of a competent military person, including the personal conviction that the military is focused on a service to the community. Likewise, motivation in the military seems driven mainly by a personal conviction about the necessity of the military mission and a responsibility for duty and service to national society. With further regard to the prediction of MTL, concrete conclusions can be drawn on certain universal value factors, military value and military virtue factors. The fact that the universal value of Materialism is not significantly associated with MTL would lead to the conclusion that the reason for pursuing a military career as a militia cadre is not primarily money-driven but rather intrinsically motivated.

Practical implications

The present data delivered new insights to reflect on the relationship between universal values and military values and virtues, and their relevance in determining OCB and MTL. Overall, this study contributed to a growing engagement of research on the importance of values and virtues in understanding the behavior of individuals at work. As Francesco and Chen (2004) stated, the understanding of individual differences in values and virtues offers broader options to cope with the diversity of employees. In a military organization, such as the Swiss Armed Forces, with an active conscriptive reserve, it is essential to motivate people who fit with their mindset in values and virtues into the military organizational culture and who are willing to perform above and beyond the call of duty. Moreover, OCB and MTL have been confirmed to be crucial for the selection of Swiss military officers (Annen et al., 2015). Accordingly, the core of this research was to show that military values and virtue factors allow for predictions regarding OCB and MTL. In accordance, the results of this study indicated that people often choose to engage in OCB because it meets certain culture-specific military values and virtues, for instance, such as represented by the factor Good Soldiership.

Generally, the results provided evidence for a first successful validation of the psycholexically derived and culture-specific military value and virtue factors. The data also supported the idea of taking military value and virtue factors into account for recruitment and selection purposes, including a measure of how an individual's values and virtues fit with the military organization.

Strengths and limitations

One advantage of this study is the combination of universal values with psycholexical and culture-sensitive military value and virtue factors to determine OCB and MTL. The results on the incremental validity of the military values and virtue factors confirmed that these factors are of high importance to be included in the examination of OCB and motivation in the military setting. However, this study had some limitations that need to be addressed in future research. First, our research design was cross-sectional, which means that even though our predictions are justified theoretically, we cannot draw causal inferences (Breckler, 1990). Future studies should apply longitudinal designs to provide more widely supported evidence for our findings. Additionally, the branch of service of the current sample of Swiss recruits was only infantry. In favor of a more representative sample, it would be helpful to include recruits from further branches of service (e.g., logistics, rescue troops) in future studies. Another limitation is the use of self-reported measures for individual-level data, which raises concerns about the possible presence of common method bias (Podsakoff, MacKenzie, Lee, & Podsakoff, 2003). By using a 5-item and 1-dimensional version to measure OCB, we neglected the fact that a measure containing different aspects of OCB might have shown different effects.

Another critical point relates to the fact that we did not examine moderator or mediator effects of potential variables in influencing the effects on OCB and MTL, as was taken into account in a few similar studies on personal values, OCB, and other work outcome (e.g., Lam et al., 2002).

Despite these limitations, relevant conclusions can be drawn from the present study. By showing that the psycholexically derived military value and military virtue factors not only determines OCB and the voluntary pursuit of a career as a militia cadre, but also showed incremental validity beyond universal values, we gained a first confirmation of the relevance of the delivered military values and virtues as assessed by the MVVC within the scope of this research. To our knowledge, no previous research has examined military values and virtues by means of a psycholexical and factor analytical approach and has verified the usefulness of the delivered factors. Regarding the effects presented in this study, further research in the military context should continue to examine the impact of military values and virtues on OCB and MTL.

General Discussion

The principle aim of the present thesis was to identify the current Swiss military core values and virtues on the basis of a psycholexical and factor analytic approach to the benefit of military leadership, training, and education.

In recent years, the psycholexical approach in combination with factor analysis has become an established way to identify and structure universal values and virtues within a variety of cultures and languages (e.g., Aavik & Allik, 2002; De Raad & Van Oudenhoven, 2008, 2011; Morales-Vives, De Raad, & Vigil-Colet, 2012, 2014; Renner, 2003b). In the military environment, the fundamental research in values and virtues is likewise grounded from a theoretical and practical point of view (Matthews, 2009; Matthews, Eid, Kelly, Bailey, & Peterson, 2006b). However, it is interesting to acknowledge that only a few studies exist which include the empirical approach to analyze the structure of military-specific values and virtues. In addition to the Pre-study and the three studies within this thesis, an international survey was conducted across 19 military organizations, which confirmed the broad relevance and the practical benefit of establishing a classification of military core values and virtues.

The following chapters summarize the main results of this research, discussing corresponding limitations, as well as theoretical and practical implications.

10 Main results and conclusions

The following outline reviews the content, goals, and findings of the four studies (Pre-study, Study I, II, and III) in reference to the six research questions.

Research question 1: Establishing a psycholexical derived catalog of Swiss military values and virtues

The first research question concerned the psycholexical identification of the military value- and virtue-describing expressions as part of the Pre-study. The Pre-study aimed at identifying a catalog of military value and virtue descriptors, being comprehensive and relevant from a military viewpoint and representing the culture of the Swiss Armed Forces. The procedure was stepwise: First, the methodological principles of the psycholexical approach were adapted to the military environment, by screening and extracting value- and virtue-describing expressions within military directives (e.g., Service Regulations 04, Instructions for officers in training, Military Ethics Report). Second, to ensure that values and virtues were approached as separate concepts, 11 military psychologists categorized the selected expressions into values and virtues according to a standardized definition by De Raad and Van Oudenhoven (2008, 2001). Third, the selected military value- and virtue-describing expressions were rated by 22 senior commanders (Brigadier General and higher) to ensure validity by the executive leadership level of the Swiss Armed Forces.

By applying a psycholexical-based search in all the relevant official military guidelines, a list of 90 military value- and virtue-describing expressions was extracted. After the categorization by military psychologists and removing 13 expressions that did not meet the criteria, the list was split into a set of 25 military values and 42 military virtues. In the last stage, the selected military values and virtues identified by psycholexical search were all confirmed to be important by the senior executive military leaders. As a result of the Pre-study, the MVVC containing 25 military values (e.g., human dignity, role model, peace, hierarchy) and 42 military virtues (e.g., punctuality, courage, integrity, loyalty) was created, and then used to capture the individual ratings in the three subsequent studies.

Research questions 2a and 2b: Assessing the structure of military values and the correlations with universal values and the Big Five of personality

Research questions 2a and 2b addressed the factorial structure of the military values, its number, and nature of factors. To answer research question 2a, one goal of Study I was to determine the factorial structure of the 25 Swiss military values as part of the MVVC. Accordingly, the military values were rated by 550 career officers and career NCOs and subjected to a principal component analysis, in combination with Goldberg's top-down approach. Five military value factors were identified, labeled Freedom (I), Social Cohesion (II), Good Soldiership (III), Mutual Respect (IV), and Military Conformity (V). An additional sixth factor was interpreted to represent a value mindset combining trust and honesty, while exhibiting reduced acceptance for multiculturalism and coexistence of minorities. Since the sixth factor was weak, and not interpretable, it was excluded as a valid factor.

Among the five military value factors, factor III Good Soldiership (containing performance of duty, performance of the mission, role model, honor, and obedience of the law), was interpreted as the only factor referring to pro-individual values (as conceptualized by Hofstede, 1997). In contrast, the other four military value factors of Freedom, Social Cohesion, Mutual Respect, and Military Conformity contained interpersonal values and reflected a pro-social character.

Referring to the three characterizing aspects of a military culture by Lang (1965; as described in section 1), the first aspect of communal character of military life is well reflected in the identified military value factors with a predominant focus on pro-social values. Furthermore, factor V Military Conformity is in line with the second aspect referring to the importance of hierarchy, formal rules, and obedience in military organizations.

The identification of the five military value factors represents the specific value culture of a militia army. More specifically, the three factors of military values "Freedom," "Social Cohesion," and "Mutual Respect" are addressing the human focus within the context of military education and leadership. The other two factors "Good Soldiership" and "Hierarchy" are pointing to the behavior in conducting military orders and in integrating to the hierarchical military organization. These five factors of values thus reflect the dual orientation of a militia army with its human and military responsibility. In conclusion, the value culture of a respectful, people-centered approach and a military dedication to duties and mission can be considered an integral part of being in a leadership role in the Swiss Armed Forces.

Research question 2b was about analyzing how the resulting factors of military values correspond with (a) the five factors of universal values (Intellectualism, Balance, Religiosity, Materialism, Conservatism; Renner, 2003b), as well as with (b) the Big Five of personality (Neuroticism, Extraversion, Openness, Aggreableness, Conscientiousness; John, Donahue, & Kentle, 1991). The correlation analysis provided the essential input to interpret the outcome of the distinct military value factors. In Study I, findings regarding the correlations between the military value factors and (a) universal values showed that Conservatism had the closest relation to the five military value factors, especially to factor II Good Soldiership. The outcome regarding the correlations to (b) the Big Five of personality pointed out that the personality factors Agreeableness and Conscientiousness shared the highest variance with the military value factors. They also confirmed the result from previous studies (De Raad & Van Oudenhoven, 2008) that these two personality traits, Agreeableness and Conscientiousness, convey aspects of positive morality.

Research questions 3a and 3b: Assessing the structure of military virtues and the correlations with the five factors of character strengths measured by the VIA-IS

Research questions 3a and 3b addressed the factorial structure of the military virtues and their relation to the five factors of character strengths. To answer research question 3a, the same procedure of methodology was applied as for research question 2a. In Study II, the 42 Swiss military virtues as part of the MVVC were rated by 270 militia officer candidates. Principal component analysis in combination with Goldberg's top-down approach delivered a result of four military virtue factors: Fortitude (I), Suitable Behavior (II), Reflection (III), and Empathy (IV). Furthermore, only factor IV Empathy directed to a pro-social perspective, while the other three factors (Fortitude, Suitable Behavior, and Reflection) included a pro-individual focus. Interesting to acknowledge is the fact that the military value factors exhibited a more pro-social aspect than the military virtue factors, while the latter included stronger pro-individual characteristics. This finding corresponded with the overall distinction between values and virtues, with values being a personal principle (Rokeach, 1973b), functioning as an internal moral compass (Hitlin & Piliavin, 2004). Given the prime aim of the Swiss Armed Forces in providing services to the general public, it was considered a positive result that the military values were predominantly of pro-social nature. Virtues, in contrast, are understood to be morally good characteristics of a person, which corresponds with the finding that the extracted four military virtue factors tended to be of pro-individual nature (cf. Hofstede, 1997).

Research question 3b combined the outcome of the four military virtue factors with the five factors of character strengths (interpersonal strengths, emotional strengths, intellectual strengths, strengths of restraint, and theological strengths) measured by the Values in Action Inventory of Strengths (VIA-IS; Peterson, Park, & Seligman, 2004). Findings from Study II showed that the emotional strengths had the strongest relation to the four military virtue factors, being in line with the result by Matthews et al. (2006b), which identified emotional strengths such as integrity, bravery, and persistence among the top character strengths in the two military samples of Royal Norwegian Naval Academy Cadets and West Point Cadets. Another substantial finding relating to research question 3b revealed that the strengths of restraint correlated significantly positive with factor II Suitable Behavior and factor IV Empathy, but negatively with factor I Fortitude. This implies the interesting aspect that strengths of restraint, depending on the situation, can be either beneficial or inhibitory in the military setting. The overall findings from Study II pointed out that the structure of military virtues in the Swiss Armed Forces partly corresponded with the structure of the universal-related character strengths.

Research question 4: Investigating the criterion validity of the military value and military virtue factors

The fourth research question pertained to the effects of the universal values, the five military value factors, and the four military virtue factors on organizational citizenship behavior (OCB) and motivation to lead (MTL). Accordingly, the purpose of Study III was first to address the question as to what extent the universal values, military values, and military virtues determine OCB (Organ, 1997) and MTL (Chan & Drasgow, 2001). OCB reflects the willingness to do more than is normally demanded; MTL represents the motivation to take over a cadre position. Second, the aim was to examine the incremental validity of the military value and military virtue factors to determine OCB and MTL, beyond the capability of the universal values. This last study was recognized as an initial validation of the military value and virtue factors, as identified in Study I and II.

Intellectualism and Harmony among the universal values, as well as the military factors Good Soldiership (MiVa-III) and Fortitude (MiVi-I) were shown to be significant determinants of OCB and MTL, as being assessed by multivariate analyses. The military value and virtue factors provided higher criterion validity in regards to OCB than to MTL. This confirmed the assumption that the motives of OCB include value- and virtue-expressive aspects, referring to a motivational behavior pattern ready to socially assist beyond the minimum expectations (Clary et al., 1998). Furthermore, the military values and military virtues proved to have an incremental contribution to the prediction of OCB and MTL beyond the one emerging through universal values, based on the outcome of a hierarchical regression analysis. More specifically, the five military value factors and the four military virtue factors explained additional variance (in comparison to the five universal value factors by Renner, 2003b) regarding criterion validity towards OCB and MTL. In summary, the outcome of the analysis for this fourth research question confirmed the validity of the resulting five military value and four military virtue factors through regression analysis in reference to the prediction of OCB and MTL.

11 Strengths and limitations

This thesis initiated a novel approach to research values and virtues in the Swiss Armed Forces. This is the first military study to identify the descriptive definition of military values and virtues by means of a psycholexical approach and to determine the underlying military value and virtue factors, thereby setting the stage to better understand the military culture in the Swiss Armed Forces.

Within the particular scope of the international survey among different military organizations (section 4.2), it was confirmed that currently none of the military organizations applied a corresponding approach of empirical research on military values and virtues. This highlighted again that one of the unique strengths of this thesis related to the empirical approach of using the psycholexical and factor analytic method to assess the structure of values and virtues within a military organization.

This approach required a wide scope of consultations, conducting personal interviews and online-questionnaires with several representative groups of the Swiss Armed Forces, including the authors of the Swiss Military Ethics Report, the highest leadership level of military generals, samples of career officers and NCOs, and militia officer candidates and recruits. To capture a wide scope of data was the prime prerequisite to conduct the factorial analysis, to apply the principal component analysis in combination with Goldberg's top-down procedure, and to identify the core values and core virtues representing the culture of the Swiss Armed Forces. The practical relevance of those values has been demonstrated with large samples of participants in the three cultures, who provided ratings on the extent to which the values were useful or important for them in their lives. More precisely, one of the strengths of this thesis relates to the inclusion of different representative samples across varying hierarchical levels. This approach in systematically collecting data points delivered a comprehensive set of records representing the individual perception of different groups of Swiss soldiers on value and virtues.

Another strength of this study design concerned the possibility to assess the congruence of values as propagated across the different organizational levels, from the higher levels of executive leaders, to the subordinated military professionals, and down to the soldier level. As mentioned above, the MVVC was administered to various types of military samples, including the high executive level (Pre-study) and the subordinated levels of career officers and career NCOs (Study I), militia cadre (Study II), and recruits (Study III) within the military hierarchy. A good quality of congruency is assumed to reflect a common and consistent base of common values and virtues, contributing to the quality in education and training. It will be subjected to additional research to further assess the congruency in values and virtues between the different organizational levels.

The question remained whether the identified values and virtues were relevant enough to be called "core" values and virtues. The answer is yes, supported by the fact that the practical relevance of the values and virtues was demonstrated on the basis of the large samples of different military subgroups, who provided ratings on the extent to which the values were important for them and the extent to which the virtues applied to them as a military person. An essential advantage of the applied research approach is that the empirical test of relevance of the values and virtues is provided by the ratings of the participants (De Raad et al., 2017) and goes beyond theoretically derived values. Furthermore, this research combined the conceptual and methodological principles across positive psychology, personality psychology, and traditional military psychology, enhancing the scope of research in the Swiss Armed Forces, with the resulting benefits as an outcome.

While pointing to strengths of the research, the limitations are being addressed likewise. In principle, a lexical-based assessment of values and virtues is a demanding and time-consuming process. It is critical to decide on a standardized definition to extract the appropriate value- and virtue-descriptors. Within this thesis, the definitions of the Dutch studies of De Raad & Van Oudenhoven (2008, 2011) were adopted. Instead of using a standardized language-specific dictionary, the corresponding expressions were identified and selected within military documentation, resulting in a so-called military psycholexical approach. Unlike the traditional dictionary, the military documentation was composed on the basis of normative directives through extractions of descriptive paragraphs (see section 6.1.1). Correspondingly, the military documentation exposed a higher degree of normative value- and virtue-describing expressions, which were desirable from a military viewpoint. Accordingly, the military value- and virtue-describing expressions received a high degree of acceptance by most of the military participants. Especially in the study with the high executive leaders (Pre-study), the ratings of relevance resulted in a right-skewed distribution, showing a tendency to a ceiling effect.

At the same time, it needs to be recognized that using the descriptive expressions of military values and virtues as part of the MVVC captures subjective ratings primarily of preferences, rather than true behavior. Albert (1956) describes this as follows: "Further, especially in the case of values, there is usually some discrepancy between what is believed and what is said and done, between what is asserted and what may be consciously or unconsciously believed" (p. 223).

Another weakness is that the different studies were subject to variation in sample quality, considering career officers versus militia officer candidates versus recruits. Accordingly, the variation in the sourced samples might impact the results from the factor analysis, not accounting for the variation in answer behavior and quality of data point. The career officers and the career NCOs have already been socialized in the organization for a long time and have adopted the military values and virtues to a greater extent. Accordingly, it can be expected that the factor structure will vary with the difference in human characteristics across the military hierarchy. In reality, the factor structure of the recruits might differentiate from the factor structure of the military personnel (career officers and NCOs). For this reason, there is a need for further studies since the factor structure of the military values and virtues was not replicated in different samples as part of this thesis. However, it is expected that the factorial structure of the military values and virtues is a timely representation and subject to change while the organization and culture of the Swiss Armed Forces is developing.

The Regulation to the Federal Staff (Swiss Confederation, 2001, Art. 7) requires an equal promotion of the three official national languages of German, French, and Italian. This makes it mandatory to extend the psycholexical search to corresponding French and Italian documents, while accounting for the differences in culture.

A further limitation referred to the use of a cross-sectional design, which especially in Study III implied that causal inferences cannot be drawn, even though the predictions are justified theoretically (Breckler, 1990).

12 Implications for research and practice

This thesis added new perspectives for extended research. The focus was on assessing the structure of military values and virtues and to transform the measured outcome into beneficial practice. The current findings illustrate how the psycholexical approach can be adapted to a military or civilian, non-military organization, allowing for the identification of culturally sensitive characteristics of the organization.

As Morales-Vives et al. (2014) pointed out, it is difficult to generalize a system of values and virtues from one culture to another. It is also not advisable to transfer questionnaires for assessing moral characteristics from one culture to another. The results found in Study I and Study II on the structure of military values and virtues provided distinct evidence that military values and virtues differ from the universal structures found in different cultures. This finding needs to be accounted for when assessing military values and virtues. In addition, the illustration of the procedure of developing the MVVC in the Pre-study may provide directive insight to other researchers in establishing successful military questionnaires to assess values and virtues. As a starting point, the focus is on the psycholexical search for expressions which distinctly represent military values and virtues, providing a framework for distinguishing how the value and virtue factors from other military organizations differ from the current research within the Swiss Armed Forces. Thereby, the results of this thesis point to a new initiative to compare different military studies applying a psycholexical approach.

Furthermore, the assessment of the value- and virtue-expressions in the survey with international military organizations (section 4.2.) revealed that in practical military use there is no consistent differentiation distinguished between the concepts of values and virtues. Within research, it is therefore suggested to treat the two concepts separately when assessing values and virtues in a military organization. This will prevent an inconsistent use of the value and virtue terms. In addition, it is essential to combine a psycholexical approach with the factor analytical assessment, to ensure the use of valid value and virtue terms and to identify cultural variances across different military organizations.

The empirical outcome of the five military value factors (Study I) and four military virtue factors (Study II) delivered the principle data required to develop a classification of military core values and virtues. Study III verified the practical importance of the identified military value and virtue factors in relation to their ability to determine OCB and MTL. The research findings highlighted the benefit of a psycholexical and factor analytic approach applied to an organizational framework with its different hierarchical levels and specific culture. This approach resulted in an evidence-based value and virtue structure, allowing for a better understanding of the specific culture of the Swiss military organization. Since a comparable psycholexical study on military values and virtues has presumably not been conducted, this research represented a generic model and approach to assess and define the core values and virtues as they currently apply to the Swiss Armed Forces. Although, in this research, value and virtue factors were captured on an individual basis, the understanding of military core values and virtues evolved on an organizational level (Pathak, Rani, & Goswami, 2016). In extension, the classification of the core values and virtues is destined to become an integral part of the military education, thereby allowing future researchers to monitor and verify the influence of the military education and training (Annen, Steiger, & Zwygart, 2004).

The key outcome of this research delivered five military core values defined as Freedom (MiVa-I[23]), Social Cohesion (MiVa-II), Good Soldiership (MiVa-III), Mutual Respect (MiVa-IV), and Military Conformity (MiVa-V), and four military core virtues defined as Fortitude (MiVi-I[24]), Suitable Behavior (MiVi-II), Reflection (MiVi-III), and Empathy (MiVi-IV).

---

23 MiVa(-I to V) = Military value factor (I to V)
24 MiVi(-I to IV) = Military virtue factor (I to IV)

It was the assumption of this research that a binding classification of military core values and virtues provides an opportunity to further shape the culture of the Swiss Armed Forces. Thus, a classification of the five military core values and the four military core virtues adds *clarity*, fosters *commitment*, and supports the *educational tasks* of the Swiss Armed Forces.

The following review outlines the resulting benefit for the Swiss Armed Forces as a result of implementing the five military core values and four military core virtues.

To provide clarity and transparency

As stated in the Swiss Report of Military Ethics (2010), the current practice is that every military commander has the right to select his or her preference in applying values and virtues. As an example, the expression *human dignity* is perceived and interpreted in many ways, e.g., as a fundamental right, as an ethical principle, as freedom to act, and as freedom to express. This illustrates the diverse degree for individual perception and interpretation of its meaning.

As part of the Pre-study, 90 expressions were identified through the psycholexical approach, which describe values and virtues. This highlighted the large numbers of expressions addressing the generic subject of values and virtues, embedded within the Swiss military documentation and guidelines (Baumann, 2007). The systematic procedure of the psycholexical approach, the inclusion of ratings by military psychologists and by the executive military leaders, allowed the researchers to define a military-specific catalog of military values and virtues (MVVC) with two separate lists of 25 military values and 42 military virtues. In a subsequent step, the factor analytic analysis enabled a further reduction of the expressions by number and complexity. Thanks to the outcome of a dedicated factor solution, it was possible to identify expressions of the same and similar content and to allocate them into factorial groups, i.e., to five military value factors and four military virtue factors. In summary, the psycholexical and factor analytic method was used to structure a long list of descriptive expressions of values and virtues into a manageable number of five military core values and four military core virtues. This represented the principle requirement to transform abstract expressions into unambiguous and clearly interpretable values and virtues. This simplifies the clarity regarding implementation as part of the military education and training (Kernic & Annen, 2016).

With regards to the concrete practical implementation, one option would be to provide each military commander the option to select his or her most personally preferred value and virtue expressions from each of the category from the five military core value factors and four military core virtue factors. As an example, the military value factor Social Cohesion (MiVa-II) and the military virtue factor Empathy (MiVi-IV) categorized into the corresponding counterparts:

- MiVa-II: comradeship, coherence, solidarity, teamwork, trust, and esprit de corps
- MiVi-IV: consideration, welfare, charity, selflessness, unselfishness, and loyalty

For a detailed overview of the military value and virtue factors, see Tab. 16 and Tab. 17 in section 9.1.1.

Above all, the definition of military core values and virtues provides guidance for adhering and promoting the corresponding clarity and transparency in regards to value and virtue education. This adds to the clear military mission, and requires that the military leaders provide clarity as to *how* to accomplish the mission and as to which values and virtues thereby to convey.

To build commitment and cohesion

The leadership role entails a certain amount of risk (Annen, 2017). In an experiment, it was shown that even those individuals who described themselves as above-average honest and socially competent changed their moral attitudes and perceptions as soon as they were placed in a leadership position (Bendahan, Zehnder, Pralong, & Antonakis, 2015). This was explained by the assumption that leaders with the appropriate power would change their moral norms of behavior, putting selfishness over the well-being of their employees, and being susceptible to corrupt practices. Military core values and virtues are understood to safeguard normative behavior and to counteract deviations thereof. Specifically, they function as a common inner moral compass in military daily operation, military education, and military training (Kernic & Annen, 2016).

To encourage personal reflection and enable ongoing evaluation of the effects of military education

It is part of the overall accountability and education of the Swiss Armed Forces to explicitly convey values and virtues. Military education is understood as a targeted influence on the values and virtues as being accepted by the Swiss Armed Forces' soldiers (Annen, Steiger, & Zwygart, 2004). This requires an obligation for the Swiss Armed Forces to regularly monitor and validate the effect of the military education and its propagation of military core values and virtues across the military organization (Eggimann & Annen, 2014). It is of particular interest to ensure that the military core values and virtues reach all military hierarchical levels and subgroups (e.g., each language region, the corps of military personnel, the corps of militia officers, and militia NCOs). In conclusion, the MVVC proved to be a valuable concept and tool to allow capturing and understanding the personal preferences for values and virtues on the basis of the underlying factorial structure. Furthermore, the MVVC enables verification of the congruence between different organizational levels with regards to preferences for values and virtues. This also encourages the military leaders to regularly apply the discipline of self-reflection and dialogue regarding their personal relation to core values and virtues.

As a preliminary conclusion, the definition and use of the identified five military core values and four military core virtues enable transparency, particularly in reference to value- and virtue-based military education. As a militia army, the Swiss Armed Forces is obliged to have a clear relation to values and virtues, both internally and externally towards society. This obligation relates to the fact that the Swiss Armed Forces is entrusted with the education of young soldiers (Annen et al., 2004).

It is therefore essential to invest in the recruitment, selection, training, and further education of executives, as they are ultimately the ones who have to credibly convey the values and virtues to the subordinates (Kernic & Annen, 2016). This is in line with the assumption that managers have a key impact on the organization (Day & Lord, 1988; Hogan, Curphy, & Horgan, 1994). The investment in selection, training, and development of executives can pay off multiple times or may cause harmful effects (Staffelbach, 2006). It was also postulated that the selection of the appropriate cadres can be made cohesive in terms of identifying candidates who share with the core values and virtues of the organization (Annen, 2017). Given the importance of executives as influencers of corporate culture (Schein, 1992) and the considerable impact that unethical leadership implies, it is appropriate to have a clear understanding of the required values and virtues for selection. In summary, it is suggested to align selection instruments with the core values and virtues. The current findings will further assist the effort to optimize the selection and training programs for Swiss soldiers at all hierarchical levels.

13 Open questions and further research

Military service is a challenge for every individual concerned. Military personnel need to be able to handle situations that occur suddenly and unpredictably, with an unknown and complex content (Boe, 2017). As Matthews (2014) stated, the current kind of conflicts, such as cyberattacks, terrorist threats, and ideological manipulations, are constantly forcing soldiers to reflect on their role, norms, and values. To foresee future risks, the nature of conflicts is predicted to become more complex and dynamic (Freedman, 2017). Rising investments in security and risk management confirm that politics is responding to increasing complexity and anticipatory needs. In November 2017, the Federal Council approved the financing of Swiss airspace protection, spending up to CHF 8 billion for the renewal of combat aircrafts and air defense systems, accounting for an annual increase of the Swiss Armed Forces budget by 2% (Federal Council, 2017).

At the same time, the effort and expertise required for the job as a military officer becomes increasingly demanding (Stocker, Jacobshagen, Semmer, & Annen, 2010). Thus, military personnel are required "to do more with less personal resources" while at the same time their appreciation of their own work is reduced (Proyer, Annen, Eggimann, Schneider, & Ruch, 2012). As a consequence, it is essential to optimize personal resources, bearing in mind that focusing on the values, virtues, and character strengths of the military personnel has been shown to enhance effective leadership and organizational climate (Matthews, 2014).

Studies to predict performance in the military and other high-risk organizations have traditionally been based on measurements of cognitive characteristics and personality traits (Elsass, Fiedler, Skop, & Hill, 2001; Picano & Roland, 2012; Picano, Williams, & Roland, 2006). As Boe (2017) emphasized, it is important to keep in mind that personality is about differences between individuals when it comes to how one reacts to circumstances, while character is about the values and virtues that govern the actions and behavior.

As Matthews, Lerner, and Annen (2019) stated with the "25/75 rule," up to 75% of variation in human performance in the military setting is accounted for by non-cognitive factors such as values, virtues, and character strengths. This is in line with findings from various military studies that character strengths, virtues, and values are regarded as highly influential on individual adaptation ability, work satisfaction, performance, and effective leadership (Matthews et al., 2006b). The significance of positive individual characteristics such as values and virtues will further increase, highlighting the need for extended research (Matthews, 2014).

Specific suggestions for research

First, several studies on values of young adults confirmed that values are subject to change over generations (e.g., 17. Shell Youth Study 2015, Youth Barometer by the Credit Suisse 2018), in line with the observation that young adults enter the army with a shift in values and virtues (Kernic & Annen, 2016). Furthermore, the current reform of the Swiss Armed Forces will be accompanied with an organizational change. Considering the principle change of the value- and virtue-based culture, it will require an integrated effort to keep the psycholexically derived set of the 25 military values and 42 military virtues up-to-date. This implies synchronizing with the ongoing renewal of contextual expressions within the MVVC as a result of changing military guidelines and personal preferences. Likewise, the inherent structure of the military values and virtues needs to be verified and adapted to the change in organizational culture.

Second, it is recommended to replicate the value and virtue structure found in the current research with additional military samples. Furthermore, the variation in military values and virtues between different military subgroups, (e.g., comparing German-, French-, and Italian-speaking persons) needs to be captured. In addition, peer-ratings (e.g., by comrades, subordinates, or commanders) on the MVVC could be included and compared with the assessment of the self-ratings (analogous to De Raad & Van Oudenhoven, 2008, 2011). Furthermore, a promising area of study would be to investigate the viewpoint of the general Swiss public regarding the values and virtues conveyed by the Swiss Armed Forces, to discover whether the core values and virtues overlap with society's viewpoint and to compare this "outside-" and "military-inside view."

Third, the sequence within the classification of the five core military values and four military virtues was not specifically analyzed yet. A pair-wise presentation to participants would be of interest, exploring the question of which value and virtue would be preferred, in comparison to each other, e.g., dignity or performance of mission. This would be in line with the theoretical assumption that values contradict and should be measured by ranking scales (Inglehart, 1977).

Fourth, due to the conceptual similarities in definition, there might be overlaps between the military value and virtue factors. For instance, MiVa-III Good Soldiership includes values (i.e., honor, performance of duty) similar to the virtues of MiVi-II Suitable Behavior (i.e., sense of honor, sense of responsibility). The military value and virtue factors emerged empirically, but they have not been substantiated in terms of reproducibility and construct validity. Thus, a further interesting question would be to analyze the convergent and discriminant validity of the five military value factors and four military virtue factors.

Fifth, the MVVC is not a valid instrument yet. Further research is needed to develop a psychometric questionnaire to identify the value and virtue profile of the soldiers based on the organizational value and virtue factor structure.

As a prerequisite, a library of behavioral scenarios must be created. This will be accomplished through an effort by military experts providing their inputs as to how the typical military values and virtues correspond with distinct behavior patterns in daily military life. Conversational sessions between military personnel will enable the identification of those behavioral aspects which represent the characteristics of the military values and virtues. The behavioral scenario catalog will allow researchers to define test items and situational behavior which can then provide elements to design a situational judgment test or a questionnaire assessing military value and virtue preferences.

Sixth, in extension to cross-sectional data it would be interesting to conduct longitudinal surveys, looking at the value and virtue change of soldiers during training sequences and measuring the effectiveness and sustainability of military education.

Seventh, it is proposed to research the correlations between further criteria such as work motivation, satisfaction with work, fulfillment, resilience or actual performance measure, and military values and virtue factors. It is likewise of interest to try to predict a new recruit's behavior, identifying values and virtues for outcome, which deviates from predicting success at other levels in the organization.

Eighth, in terms of a top-down verification it would be interesting to verify how the ratings by the military executive leaders from the Pre-study compare with the ratings by the military professionals, and the subordinate leadership levels.

Ninth, this thesis had its primary focus on the Swiss military values and virtues. In extension, the international study on the practical usage of core values and virtues implied that the military's view on values and virtues is not consciously differentiating between values and virtues. Accordingly, it would be of interest to extend the initial consultation with military organizations outside of Switzerland and to achieve a common understanding regarding the conceptual differences of values and virtues. Furthermore, findings regarding values and virtues and its underlying structure cannot simply be transferred from one language to the other (cf. Berry, 1969). For the purpose of cross-cultural comparability, it is essential to apply similar psycholexical procedures in studying the factorial structures of values and virtues in various military organizations, depending on their different combat or training missions. The assumption by Matthews (2008) that the values, virtues, and character strengths essential in combat differ from those vital to success in training or in an administrative job within the military, has not yet been empirically tested and would be an interesting future research question.

14 Final comments

Within the framework of positive psychology, a military organization represents a positive institution (Matthews, 2008). This underlines the significance of the Swiss Armed Forces to be seen as a role model from the perspective of organizations and society as a whole. Accordingly, this research project provided a contribution to the educational mission of the Swiss Armed Forces, as introduced by General Wille (1954), based on a systematic assessment of the current Swiss military value- and virtue-based culture.

Upon receiving a special award from the U.S. Military Academy, Condolezza Rice (2014) delivered one of her core statements: "Nothing of value is ever won without sacrifice." This focus relates back to the mindset introduced in the USMA Cadet Prayer, quoted at the beginning of this thesis as initially referred to in this thesis: "Make us to choose the harder right instead of the easier wrong and never to be content with a half truth when the whole can be won" (U.S. Military Academy, 2017).

Thus, the claim of a commitment to values and virtues cannot be accomplished without the essential investment. This implies the effort to create a classification of military core values and virtues and to pursue a consistent selection, education, and development of soldiers. The content of this thesis explored the scope of capturing the relevant data, to analyze and deliver the military core values and core virtues, based on the psycholexical and factor analytic definition as it applies to the Swiss Armed Forces.

A classification of core values and core virtues as well as its practical implementation can also be seen as the representation of an organization, similar to a company business card. It reflects transparency and willingness for regular self-reflection and evidence-based verification.

Defining and recognizing the organization's core values and virtues establish a firm commitment towards preserving a military organizational culture, and ensuring that it is in line with the principles of positive psychology.

List of Figures

Theoretical Background

Fig. 1: Summary of the Pre-study and the three studies within the Swiss Armed Forces 37

Fig. 2: Theoretical model of relations among the ten motivational types of value as proposed by Schwartz (2006) (illustration drawing from Schwartz, 2006) 64

Fig. 3: Structure of the MVVC 117

Study I

Fig. 4: The emergence of factors from the 25 military value descriptors (first unrotated principal component) starting from a general factor to six-factor solution rotated according to the varimax criterion ($N=550$); numbers within boxes indicate the number of factors extracted for a given level. Correlation coefficients to adjacent factors are only displayed when exceeding a coefficient > .20. Boldface indicates final five-factor solution 130

Study II

Fig. 5: The emergence of factors from the 36 military value descriptors (first unrotated principal component) starting from a general factor to five-factor solution rotated according to the varimax criterion ($N=270$); numbers within boxes indicate the number of factors extracted for a given level. Correlation coefficients to adjacent factors are only displayed when exceeding a coefficient >.20. Boldface indicates final four-factor solution 152

List of Tables

Theoretical Background

Tab. 1: Classification of the six core virtues and 24 character strengths (Peterson & Seligman, 2004, pp. 29–30) 43
Tab. 2: Values of the RVS (Rokeach, 1973a) 61
Tab. 3: Definitions of the ten value types by Schwartz (1992), cited according to Mohler and Wohn (2005) 62
Tab. 4: Overview of the systems of values (adapted to the version of Morales-Vives et al., 2012) 83
Tab. 5: Overview of the systems of virtues (adapted to the version of Morales-Vives et al., 2014) 86 87
Tab. 6: Overview of the core values and virtues as collected in the international study on values and virtues in military organizations ... 100
Tab. 7: List of main sources on Swiss military values and virtues used for the psycholexical search (adapted from the Swiss Federal Report on Military Ethics, 2010, p. 24) 110
Tab. 8: Descriptive statistics of 25 military value descriptors from the perspective of the Swiss military Generals 113
Tab. 9: Descriptive statistics of 42 military virtue descriptors from the perspective of the Swiss military Generals 115

Study I

Tab. 10: Descriptive statistics of the 25 military value descriptors 129
Tab. 11: Varimax loadings of the five factors based on the ratings of the 25 military value descriptors 132
Tab. 12: Partial correlations (controlled for age) of the factor scores of the five-factor solution with the five universal value types (AVQ; Renner, Salem, & Alexandrowicz, 2004) and the five factors of personality (BFI; John, Donahue, & Kentle, 1991) 134

Study II

Tab. 13: Descriptive statistics of 42 military virtue descriptors 150
Tab. 14: Varimax loadings of the four military virtue factors based on the ratings of 36 military virtue descriptors 153
Tab. 15: Correlations between four military virtue factors and five factors of character strengths (VIA-IS; Peterson, Park, & Seligman, 2005) 155

Study III

Tab. 16: Item contents with the loadings on the five factors of military values as identified by the principal component analysis 163

Tab. 17: Scale content with the loadings on the four factors of military virtues as identified by the principal component analysis 164

Tab. 18: Means, standard deviations, and correlations among the study variables in the sample of Swiss recruits 172

Tab. 19: Summary of multiple regression analysis for universal values (AVQ; Renner, Salem, & Alexandrowicz, 2004), military values, and military virtues determining organizational citizenship behavior (OCB; Meierhans, Rietmann, & Jonas, 2008) and motivation to lead (MTL; Swiss Armed Forces, 2012) 174

Tab. 20: Summary of hierarchical regression analysis for universal values, military values, and military virtues determining organizational citizenship behavior (OCB; Meierhans, Rietmann, & Jonas, 2008) and motivation to lead (MTL; Swiss Armed Forces, 2012) 175

References

Aavik, T., & Allik, J. (2002). The structure of Estonian personal values: A lexical approach. *European Journal of Personality, 16,* 221–235. doi:10.1002/per.439.

Ajzen, I. (1991). The theory of planned behavior. *Journal of Studies on Alcohol and Drugs, 72*(2), 322–332. doi:10.1080/10810730.2011.551991.

Ajzen, I., & Fishbein, M. (1977). Attitude-behavior relations: A theoretical analysis and review of the research. *Psychologia, 84,* 888–918. doi:10.1037/0033-2909.84.5.888.

Albert, E. M. (1956). The classification of values: A method and illustration. *American Anthropologist, 58*(1), 403–409. doi:10.1525/aa.1956.58.2.02a00020.

Algoe, S. B., & Haidt, J. (2009). Witnessing excellence in action: The "other-praising" emotions of elevation, gratitude, and admiration. *Journal of Positive Psychology, 4,* 105–127. doi:10.1080/17439760802650519.

Allport, F. H., & Allport, G. W. (1921). Personality traits: Their classification and measurement. *Journal of Abnormal and Social Psychology, 16,* 6–40.

Allport, G. W. (1937). *Personality: A psychological interpretation.* New York, NY: Holt.

Allport, G. W., & Vernon, P. E. (1931). *A study of values.* Oxford, UK: Houghton Mifflin.

Aluja, A., & García, L. F. (2004). Relationships between Big Five personality factors and values. *Social Behavior and Personality: An International Journal, 32*(7), 619–625. doi:10.2224/sbp.2004.32.7.619.

American Psychiatric Association. (2000). *Diagnostic and statistical manual of mental disorders* (4th ed., text rev.). Washington, DC: American Psychiatric Association.

Ang, S., Van Dyne, L., & Begley, T. M. (2003). The employment relationships of foreign workers versus local employees: A field study of organizational justice, job satisfaction, performance, and OCB. *Journal of Organizational Behavior, 24*(SPEC. ISS.), 561–583. doi:10.1002/job.202.

Angleitner, A., Ostendorf, F., & John, O. (1990). Towards a taxonomy of personality descriptors in German: A psycho-lexical study. *European Journal of Personality, 4*(March), 89–118. doi:10.1002/per.2410040204.

Annen, H. (2017). Kaderselektion als Mittel zur Gestaltung der Führungskultur [Cadre selection as a means of shaping the leadership culture]. *Military Power Review, 2,* 36–49.

Annen, H., Goldammer, P., & Szvircsev Tresch, T. (2015). Longitudinal effects of OCB on cadre selection and pursuing a career as militia cadre in the Swiss Armed Forces. *Military Psychology, 27*(1), 9–21. doi:10.1037/mil0000063.

Annen, H., Steiger, R., & Zwygart, U. (2004). *Gemeinsam zum Ziel: Anregungen für Führungskräfte einer modernen Armee [Heading together to target: Implications for leaders of a modern army]*. Bern, Switzerland: Huber.

Apelt, M. (2010). Militärische Sozialisation [Military socialization]. In N. Leonhard & I.-J. Werkner (eds.), *Militärsoziologie – Eine Einführung [Military sociology – an introduction]*, 428–444. Wiesbaden, Germany: VS Verlag für Sozialwissenschaften.

Aquinas, S. T. (1989). *Summa theologiae* (translated by T. McDermott). Westminster, MD: Christian Classics.

Aristotle. (1984). *The complete works of Aristotle* (edited by J. Barnes). Princeton, NJ: Princeton University Press.

Aristotle. (2000). *Nicomachean ethics* (translated by R. Crip). Cambridge, UK: Cambridge University Press.

Arjoon, S. (2000). Virtue theory as a dynamic theory of business. *Ethics, 28*(1997), 159–178. doi:10.1023/A:1006339112331.

Arjoon, S. (2008). Reconciling situational social psychology with virtue ethics. *International Journal of Management Reviews, 10*(3), 221–243. doi:10.1111/j.1468-2370.2007.00216.x.

Aronovitch, H. (2001). Good soldiers: A traditional approach. *Journal of Applied Philosophy, 18*(1), 13–23.

Asendorpf, J. (2004). *Psychologie der Persönlichkeit [Psychology and personality]* (3rd ed.). Berlin, Germany: Springer.

Athota, V. S., & O'Connor, P. J. (2014). How approach and avoidance constructs of personality and trait emotional intelligence predict core human values. *Learning and Individual Differences, 31*, 51–58. doi:10.1016/j.lindif.2013.12.009.

Bachman, J. G., Sigelman, L., & Diamond, G. (1987). Self-selection, socialization, and distinctive military values: Attitudes of high school seniors. *Armed Forces & Society, 13*(2), 169–187. doi:10.1177/0095327X8701300201.

Bakan, D. (1966). *The duality of human existence*. Chicago, IL: Rand McNally.

Ball-Rokeach, S. J., & Loges, W. E. (1996). Making choices: Media roles in the construction of value-choices. In C. Seligman & J. M. Olson (eds.), *The psychology of values: The Ontario symposium* (pp. 277–298). Hillsdale, NJ: Lawrence Erlbaum.

Banth, S., & Singh, P. (2011). Positive character strengths in middle-rung army officers and managers in civilian sector. *Journal of the Indian Academy of Applied Psychology, 3*, 320–324.

Bardi, A., & Schwartz, S. H. (2003). Values and behavior: Strength and structure of relations. *Personality and Social Psychology Bulletin, 29*(10), 1207–1220. doi:10.1177/0146167203254602.

Baron, H. (1996). Strengths and limitations of ipsative measurement. *Journal of Occupational and Organizational Psychology, 69*(1), 49–56. doi:10.1111/j.2044-8325.1996.tb00599.x.

Battistelli, F., Ammendola, T., & Galantino, M. G. (1999). The fuzzy environment and postmodern soldiers: The motivations of the Italian contingent in Bosnia. *Small Wars & Insurgencies, 10*(2), 138–160. doi:10.1080/09592319908423244.

Baumann, D. (2007). Militärethik Armee [Military ethics]. *Theologie und Frieden [Theology and peace]*. Stuttgart, Germany: Kohlhammer.

Baumgartner, E., & Reimherr, A. (2006). *Essays über Carl Stumpf und Franz Brentano [Essays about Carl Stumpf and Franz Brentano]*. Würzburg, Germany: Internationale Franz Brentano Gesellschaft.

Bendahan, S., Zehnder, Ch., Pralong, F. P., & Antonakis, J. (2015). Leader corruption depends on power and testosterone. *The Leadership Quarterly, 26*, 101–122. doi: 10.1016/j.leaqua.2014.07.010.

Berings, D., De Fruyt, F., & Bouwen, R. (2004). Work values and personality traits as predictors of enterprising and social vocational interests. *Personality and Individual Differences, 36*(2), 349–364. doi:10.1016/S0191-8869(03)00101-6.

Berkowitz, M. W. (2002). The science of character education. In W. Damon (Ed.), *Bringing in a new era in character education* (pp. 43–63). Stanford, CA: Hoover Institute Press.

Berry, J. W. (1969). On cross-cultural comparability. *International Journal of Psychology*, 119–128. doi:10.1080/00207596908247261.

Bilsky, W. (1998). *Values and motives*. Paper presented at the International Research Workshop "Values: Psychological Structure, Behavioral Outcomes, and Inter-Generational Transmission", Maale-Hachamisha, Israel (January 12-16, 1998). Münster, Germany: University, Psychological Institute.

Bilsky, W. (2005). Werte und Werthaltungen [Values and value orientations]. In H. Weber & T. Rammsayer (eds.), *Handbuch der Persönlichkeitspsychologie und Differentiellen Psychologie [Manual of personality psychology and differential psychology]* (pp. 298–304). Göttingen, Germany: Hogrefe.

Bilsky, W., & Schwartz, S. H. (1994). Values and personality. *European Journal of Personality, 8*, 163–181. doi:10.1002/per.2410080303.

Boe, O. (2017). Character strengths and their relevance for military officers. In S. Rawat (Ed.), *Military psychology: International perspectives* (pp. 113–132). Jaipur, India: Rawat Publications.

Bollnow, O. F. (1958). *Wesen und Wandel der Tugenden [Nature and change of virtues]*. Frankfurt am Main, Germany: Ullstein Taschenbücher-Verlag GmbH.

Booth, B., Segal, M. W., Bell, D. B., Martin, J. A., Ender, M. G., Rohall, D. E., & Nelson, J. (2007). *What we know about Army families: 2007 update*. Washington, DC: ICF International.

Borman, W. C., & Motowidlo, S. M. (1993). Expanding the criterion domain to include elements of contextual performance. In N. Schmitt & W. C. Borman (eds.), *Personnel selection in organizations* (pp. 71–98). San Francisco, CA: Jossey-Bass.

Borman, W. C., & Penner, L. A. (2001). Citizenship performance: Its nature, antecedents, and motives. In B. W. Roberts & R. Hogan (eds.), *Personality psychology in the workplace* (pp. 45–61). Washington, DC: American Psychological Association.

Böhm, G. (2008). Introduction to the special issue: Intuition and affect in risk perception and decision making. *Judgment and Decision Making, 3*(1), 1.

Braithwaite, V. A., & Law, H. G. (1985). Structure of human values: Testing the adequacy of the Rokeach Value Survey. *Journal of Personality and Social Psychology, 49*(1), 250–263. doi:10.1037/0022-3514.49.1.250.

Braithwaite, V. A., & Scott, W. A. (1991). Values. In J. P. Robinson & P. R. Shaver (eds.), *Measures of personality and social psychological attitudes* (pp. 661–749). San Diego, CA: Academic.

Breckler, S. J. (1990). Applications of covariance structure modeling in psychology: Cause for concern? *Psychological Bulletin, 107*(2), 260–273. doi:10.1037/0033-2909.107.2.260.

Brentano, F. (1889). *Vom Ursprung sittlicher Erkenntnis [From the origin of moral knowledge]*. Berlin, Germany: Springer.

British Army. (n.d.). Army pride: values and standards. Retrieved August 26, 2016, from http://www.army.mod.uk/join/38093.aspx#38093.

Britt, T. W., Adler, A. B., & Castro, C. A. (2006). *Military life: The psychology of serving in peace and combat, Vol. 4, military culture*. Westpoint, CT: Praeger Security International.

Britt, T. W., Stetz, M. C., & Bliese, P. D. (2004). Work-relevant values strengthen the stressor-strain relation in elite army units. *Military Psychology, 16*(1), 1–17. doi:10.1207/s15327876mp1601_1.

Brokken, F. B. (1978). *The language of personality*. Meppel, The Netherlands: Krips.

Brooks, J. E. (2010). *Midshipman character strengths and virtues in relation to leadership and daily stress and coping*. Unpublished doctoral dissertation, Washington, DC, USA: Howard University.

Brown, A. (2010). Doing less but getting more: Improving forced-choice measures with Item Response Theory. *Assessment and Development Matters*, 2(1), 21–25.

Callina, K. S., Ryanb, D., Murray, E. D., Colby, A., Damon, W., Matthews, M., & Lerner, R. M. (2017). Developing leaders of character at the united states military academy: A relational developmental systems analysis. *Journal of College and Character*, 18(1), 9–27. doi:10.1080/2194587X.2016.1260475.

Cameron, K. (2011). Responsible leadership as virtuous leadership. *Journal of Business Ethics*, 98, 25–35. doi:10.1007/s10551-011-1023-6.

Campbell, J. B., Jayawickreme, E., & Hanson, E. J. (2015). Measures of values and moral personality. *Measures of Personality and Social Psychological Constructs*, 505–529. doi:10.1016/B978-0-12-386915-9.00018-8.

Caprara, G., Vecchione, M., & Schwartz, S. H. (2009). Mediational role of values in linking personality traits to political orientation. *Asian Journal of Social Psychology*, 12, 82–84. doi:10.1111/j.1467-839X.2009.01274.x.

Carifio, J., & Perla, R. J. (2007). Ten common misunderstandings, misconceptions, persistent myths and urban legends about Likert scales and Likert response formats and their antidotes. *Journal of Social Sciences*, 3(3), 106–116. doi:10.3844/jssp.2007.106.116.

Carver, C. S., & Scheier, M. F. (1992). *Perspectives on personality*. Boston, MA: Allyn and Bacon.

Casey Jr., G. W. (2011). Comprehensive soldier fitness: A vision for psychological resilience in the U.S. Army. *American Psychologist*, 66(1), 1–3. doi:10.1037/a0021930.

Cawley, M. J., Martin, J. E., & Johnson, J. A. (2000). A virtues approach to personality. *Personality and Individual Differences*, 28(5), 997–1013. doi:10.1016/S0191-8869(99)00207-X.

Chan, K. Y., & Drasgow, F. (2001). Toward a theory of individual differences and leadership: Understanding the motivation to lead. *Journal of Applied Psychology*, 86, 481–498. doi:10.1037/0021-9010.86.3.481.

Chatman, J. A. (1991). Matching people and organizations: Selection and socialization in public accounting firms. *Administrative Science Quarterly*, 36(3), 459–484. doi:10.2307/2393204.

Cheung, M. W.-L., & Chan, W. (2002). The effects of item parceling on goodness-of-fit and parameter estimate bias in structural equation modeling. *Structural Equation Modeling: A Multidisciplinary Journal, 9*(1), 55–77. doi:10.1207/S15328007SEM0901.

Chisholm, R. M. (1986). *Brentano and intrinsic value.* New York, NY: Cambridge University Press.

Clary, E. G., Snyder, M., Ridge, R. D., Copeland, J., Stukas, A. A., Haugen, J., & Miene, P. (1998). Understanding and assessing the motivations of volunteers: A functional approach. *Journal of Personality and Social Psychology, 74*(6), 1516–1530. doi:10.1037/0022-3514.74.6.1516.

Clemans, W. V. (1966). An analytical and empirical examination of some properties of ipsative measures. *Psychometric Monograph of the Psychometric Society, 14.*

Clemmons, A. B., & Fields, D. (2011). Values as determinants of the motivation to lead. *Military Psychology, 23*(6), 587–600. doi:10.1080/08995605.2011.616787.

Closs, S. J. (1996). On the factoring and interpretation of ipsative data. *Journal of Occupational and Organizational Psychology, 69*(4), 1–47.

Collins, J. J. (1998). The complex context of American military culture: A practitioner's view. *The Washington Quarterly V, 21,* 213–228. doi:10.1080/01636609809550359.

Cornum, R., Matthews, M. D., & Seligman, M. E. P. (2011). Comprehensive soldier fitness: Building resilience in a challenging institutional context. *American Psychologist, 66*(1), 4–9. doi:10.1037/a0021420.

Corr, P. J., & Matthews, G. (2009). *The Cambridge handbook of personality psychology.* New York, NY: Cambridge University Press.

Cosentino, A. C., & Solano, A. C. (2012). Character strengths: A study of Argentina soldiers. *The Spanish Journal of Psychology, 15*(1), 199–215. doi:10.5209/rev_SJOP.2012.v15.n1.37310.

Credit Suisse. (2018). *2018 Youth Barometer: A generation under economic and professional pressure.* Retrieved September 29, 2018, from https://www.credit-suisse.com/corporate/en/responsibility/dialogue/youth-barometer.html.

Crețu, R. Z., Burcas, S., & Negovan, V. (2012). A psycho-lexical approach to the structure of values for Romanian population. *Procedia – Social and Behavioral Sciences, 33,* 458–462. doi:10.1016/j.sbspro.2012.01.163.

Crossan, M., Mazutis, D., & Seijts, G. (2013). In search of virtue: The role of virtues, values and character strengths in ethical decision making. *Journal of Business Ethics, 113*(4), 567–581. doi:10.1007/s10551-013-1680-8.

Dahlsgaard, K., Peterson, C. M., & Seligman, M. E. P. (2005). Shared virtue: The convergence of valued human strengths across culture and history. *Review of General Psychology, 9*(3), 203–213. doi:10.1037/1089-2680.9.3.203.

Davenport, M. M. (1986). The military virtues: From Aristotle to Skinner. *Southwest Philosophy Review, 3,* 161–177.

Day, D. V. & Lord, R. G. (1988). Executive leadership and organizational performance: Suggestions for a new theory and methodology. *Journal of Management, 14,* 453–464. doi:10.1177/014920638801400308.

De Raad, B. (1995). The psycholexical approach to the structure of interpersonal traits. *European Journal of Personality, 9,* 89–102. doi:10.1002/per.2410090203.

De Raad, B. (2000). *The Big Five personality factors: The psycholexical approach to personality.* Ashland, OH: Hogrefe & Huber.

De Raad, B., & Hendriks, A. A. (1997). A psycholexical route to content coverage in personality assessment. *European Journal of Psychological Assessment, 13,* S85. doi:10.1027/1015-5759.13.2.85.

De Raad, B., & Mlačić, B. (2015). The lexical foundation of the Big Five-factor model. In T. A. Widiger (Ed.), *The Oxford handbook of the five factor model* (Vol. 1). doi:10.1093/oxfordhb/9780199352487.013.12.

De Raad, B., & Van Oudenhoven, J. P. (2008). Factors of values in the Dutch language and their relationship to factors of personality. *European Journal of Personality, 22,* 81–108. doi:10.1002/per.667.

De Raad, B., & Van Oudenhoven, J. P. (2011). A psycholexical study of virtues in the Dutch language, and relations between virtues and personality. *European Journal of Personality, 25,* 43–52. doi:10.1002/per.777.

De Raad, B., Morales-Vives, F., Barelds, D. P. H., Van Oudenhoven, J. P., Renner W., & Timmerman, M. E. (2016). Values in a cross-cultural triangle: A comparison of value taxonomies in the Netherlands, Austria, and Spain. *Journal of Cross-Cultural Psychology, 47,* 1053–1075. doi: 10.1177/0022022116659698.

De Raad, B., Timmerman, M. E., Morales-Vives, F., Renner, W., Barelds, D. P. H., & Van Oudenhoven, J. P. (2017). The psycho-lexical approach in exploring the field of values: A reply to Schwartz. *Journal of Cross-Cultural Psychology, 48*(3), 444–451. doi: 10.1177/0022022117692677.

DeYoung, C. G. (2006). Higher-order factors of the Big Five in a multi-informant sample. *Journal of Personality and Social Psychology, 91,* 1138–1151. doi:10.1037/0022-3514.91.6.1138.

Digman, J. M. (1997). Higher order factors of the Big Five. *Journal of Personality and Social Psychology, 73,* 1246–1256. doi:10.1037/0022-3514.73.6.1246.

Dollinger, S. J., Leong, F. T. L., & Ulicni, S. K. (1996). On traits and values: With special reference to openness to experience. *Journal of Research in Personality, 30*(1), 23–41. doi:10.1006/jrpe.1996.0002.

Druckman, D., Stern, P. C., Diehl, P., Fetherston, A. B., Johansen, W. D., & Ratner, S. (1997). Evaluating peacekeeping missions. *International Studies, 3*(1), 151–165. doi:10.2307/222819.

Duckworth, A. L., Peterson, C. M., Matthews, M. D., & Kelly, D. R. (2007). Grit: Perseverance and passion for long-term goals. *Journal of Personality and Social Psychology, 92*(6), 1087–1101. doi:10.1037/0022-3514.92.6.1087.

Duh, M., Belak, J., & Milfelner, B. (2010). Core values, culture and ethical climate of ethical elements as constitutional between differences behaviour: Exploring and non-family enterprises. *Journal of Business Ethics, 97*(3), 473–489. doi:10.1007/sl0551-010-0519-9.

Eggimann, N., & Annen, H. (2014). Von Auftragserfüllung bis Zivilcourage – eine evidenzbasierte Werteklassifikation [From performance of mission to moral courage: An evidence-based classification of values]. *Allgemeine Schweizerische Militärzeitschrift (ASMZ), 5,* 50–51.

Eggimann, N. & Schneider, A. (2008). *Positive Psychologie und Arbeitszufriedenheit: Eine vergleichende Studie mit Berufsoffizieren und Orchestermusikern* [Positive psychology and satisfaction with work: A comparative study with career officers and orchestra musicians]. Unpublished master thesis, Zurich, Switzerland: University of Zurich.

Eid, J., Matthews, M. D., & Johnsen, B. H. (2004). Human strengths and adaptation to a radically changed context. Paper presented at the 12th Annual Convention of the American Psychological Association, Honolulu, HI.

Elizur, D., & Koslowsky, M. (2001). Values and organizational commitment. *International Journal of Manpower, 22*(7), 593–599. doi:10.1108/01437720110408967.

Elsass, W. P., Fiedler, E., Skop, B., & Hill, H. (2001). Susceptibility to maladaptive responses to stress in basic military training based on variants of temperament and character. *Military Medicine, 166*(10), 884–888.

Emmons, R. A. (1989). The personal striving approach to personality. In L. A. Pervin (Ed.), *Goal concepts in personality and social psychology.* Hillsdale, NJ: Lawrence Erlbaum.

Erikson, E. H. (1963). *Childhood and society.* New York, NY: Norton.

Erikson, E. H. (1982). *The life cycle completed.* New York, NY: Norton.

Eysenck, H. J. (1954). *The psychology of politics.* London, UK: Routledge.

Feather, N. T. (1986). Value systems across cultures: Australia and China. *International Journal of Psychology, 21*, 697–715.

Feather, N. T., & Peay, E. R. (1975). The structure of terminal and instrumental values: Dimensions and clusters. *Australian Journal of Psychology, 27*, 151–164.

Federal Council. (2017). *Funding renewed for airspace protection – Federal Council makes strategic decisions* [press release, 2017, November 8]. Retrieved December 20, 2017, from https://www.admin.ch/gov/en/start/documentation/media-releases.msg-id-68721.html.

Fischer, R., & Smith, P. B. (2006). Who cares about justice? The moderating effect of values on the link between organisational justice and work behaviour. *Applied Psychology, 55*(4), 541–562. doi:10.1111/j.1464-0597.2006.00243.x.

Fowers, B. J., & Davidov, B. J. (2006). The virtue of multiculturalism: Personal transformation, character, and openness to the other. *American Psychologist, 61*, 581–594. doi:10.1037/0003-066X.61.6.581.

Francesco, A. M., & Chen, Z. X. (2004). Collectivism in action. *Group and Organization Management, 29*(4), 425–423. doi:10.1177/1059601103257423.

Franke, V. C. (2001). Generation X and the military: A comparison of attitudes and values between West Point cadets and college students. *Journal of Political & Military Sociology, 29*(1), 92–119.

Franke, V., & Heinecken, L. (2001). Adjusting to peace: Military values in a cross-national comparison. *Armed Forces and Society, 27*(4), 567–595. doi:10.1177/0095327X0102700404.

Freedman, L. (2017). *Future of war*. London, UK: Allen Lane.

French, E. G., & Ernest, R. R. (1955). The relationship between authoritarianism and acceptance of military ideology. *Journal of Psychology, 24*, 181–191. doi:10.1111/j.1467-6494.1955.tb01183.x.

Furnham, A., Petrides, K. V., Tsaousis, I., Pappas, K., & Garrod, D. (2005). A cross-cultural investigation into the relationships between personality traits and work values. *Journal of Psychology: Interdisciplinary and Applied, 139*(1), 5–32. doi:10.3200/JRLP.139.1.5-32.

Gable, S. L., & Haidt, J. (2005). What (and why) is positive psychology? *Review of General Psychology, 9*, 103–110. doi:10.1037/1089-2680.9.2.103.

Gaylin, N. L. (1989). Ipsative measures: In search of paradigmatic change and a science of subjectivity. *Person-Centered Review, 4*(4), 429–445.

Gayton, S. D., & Kehoe, E. J. (2015). A prospective study of character strengths as predictors of selection into the Australian Army Special Force. *Military Medicine, 180*(2), 151–157. doi:10.7205/MILMED-D-14-00181.

George, J. M., & Brief, A. P. (1992). Feeling good-doing good: A conceptual analysis of the mood at work-organizational spontaneity relationship. *Psychological Bulletin, 112*(2), 310–329. doi:10.1037/0033-2909.112.2.310.

Goldberg, L. R. (1981). Language and individual differences: The search for universals in personality lexicons. *Review of Personality and Social Psychology, 2*(1), 141–165.

Goldberg, L. R. (2006). Doing it all bass-ackwards: The development of hierarchical factor structures from the top down. *Journal of Research in Personality, 40,* 347–358. doi:10.1016/j.jrp.2006.01.001.

Gowri, A. (2007). On corporate virtue. *Journal of Business Ethics, 70,* 391–400. doi:10.1007/s10551-006-9117-2.

Graumann, C. F., & Willig, R. (1983). Wert, Wertung, Werthaltung [Value, evaluation and value orientation]. In H. Thomae (Ed.), *Theorien und Formen der Motivation [Theories and types of motivation]* (pp. 312–396). Göttingen, Germany: Hogrefe.

Green, E. G. T., & Paez, D. (2005). Variation of individualism and collectivism within and between 20 countries: A typological analysis. *Journal of Cross-Cultural Psychology, 36*(3), 321–339. doi:10.1177/0022022104273654.

Greenfield, P. (2000). Three approaches to the psychology of culture: Where do they come from? Where can they go? *Asian Journal of Social Psychology, 3,* 223–240. doi:10.1111/1467-839X.00066.

Gregory, B. T., Harris, S. G., Armenakis, A. A., & Shook, C. L. (2009). Organizational culture and effectiveness: A study of values, attitudes, and organizational outcomes. *Journal of Business Research, 62*(7), 673–679. doi:10.1016/j.jbusres.2008.05.021.

Grojean, M. W., & Thomas, J. L. (2006). From values to performance: It's the journey that changes the traveler. In T. W. Britt, A. B. Adler, & C. A. Castro (eds.), *Military life: The psychology of serving in peace and combat (Vol. 4,* pp. 35–59). Westport, CT: Praeger Security International.

Guadagnoli, E., & Velicer, W. F. (1988). Relation to sample size to the stability of component patterns. *Psychological Bulletin, 103*(2), 265–275. doi:10.1037/0033-2909.103.2.265.

Guilford, J. P. (1959). *Personality.* New York, NY: McGraw-Hill.

Halbesleben, J. R. B., Bolino, M. C., Bowler, W. M., & Turnley, W. H. (2010). Organizational concern, prosocial values, or impression management ? How supervisors attribute motives to organizational citizenship behavior 1. *Journal of Applied Social Psychology, 40*(6), 1450–1489. doi:10.1111/j.1559-1816.2010.00625.x.

Haltiner, K. W. (1996). Das Militär im Wandel der Wertvorstellungen [The military in the process of changing values]. In L. E. Carrel (ed.), *Schweizer Armee: Heute und in Zukunft* [Swiss Armed Forces: Today and Future]. Thun, Switzerland: Ott Verlag.

Hampson, S. E., Goldberg, L. R., & John, O. P. (1987). Category-breadth and social- desirability values for 573 personality terms. *European Journal of Personality, 1*, 241–258. doi:10.1002/per.2410010405.

Hansen, K. P. (2000). *Kultur und Kulturwissenschaft. Eine Einführung [Culture and cultural sciences. An introduction]* (2nd ed.). Basel, Switzerland: Francke.

Hicks, L. E. (1970). Some properties of ipsative, normative, and forced-choice normative measures. *Psychological Bulletin, 74*, 167–184.

Hillen, J. (1999). The future of American military culture: Must U.S. military culture reform? *Orbis*, 43–57. doi:10.1016/S0030-4387(99)80056-8.

Hitlin, S., & Piliavin, J. A. (2004). Values: Reviving a dormant concept. *Annual Review of Sociology, 30*, 359–393. doi:10.1146/annurev.soc.30.012703.110640.

Hofstede, G. (1980). *Culture's consequences: International differences in work-related values*. London, UK: Sage.

Hofstede, G. (1997). *Cultures and organizations. Software of the mind.* New York, NY: McGraw-Hill.

Hofstede, G. (2001). *Culture's consequences: Comparing values, behaviors, and organizations across nations* (2nd ed.). London, UK: Thousand Oaks.

Hofstede, G., & Bond, M. (1984). Hofstede's culture dimensions. *Journal of Cross-Cultural Psychology, 15*(4), 417–433. doi:10.1177/0022002184015004003.

Hogan, R., Curphy, G. J., & Hogan, J. (1994). What we know about leadership effectiveness and personality. *American Psychologist, 49*, 493–504. doi:10.1037/0003-066X.49.6.493.

Horn, J. L. (1965). A rationale and test for the number of factors in factor analysis. *Psychometrika, 30*, 179–185.

Huntington, S. P. (1957). *The Soldier and the state: The theory of politics and civil-military relations*, Cambridge, MA: Belknap Press.

Hügli, A. & Lübcke, P. (2003). Wertphilosophie [Philosophy of values]. In S. Räss (ed.), *Philosophie-Lexikon [Philosophy lexicon]* (p. 672). Hamburg, Germany: Rowohlt.

Inagaki, O. (1975). The Jieitai: military values in a pacifist society. *Japan Interpreter, 10*, 1–15.

Inglehart, R. (1977). *The silent revolution: Changing values and political styles among western publics*. Princeton, NJ: Princeton University Press.

International Military Testing Association (IMTA). (n.d.). Retrieved from www.imta.info. ISI Web of Science citation report (2017, December 5). Retrieved from https://apps.webofknowledge.com/WOS_GeneralSearch_input.do?product=WOS&search_mode=GeneralSearch&SID=S1dC45okgrVAfjANeTM&preferencesSaved=.

Jackson, J. J., Thoemmes, F., Jonkmann, K., Ludtke, O., & Trautwein, U. (2012). Military training and personality trait development: Does the military make the man, or does the man make the military? *Psychological Science, 23*(3), 270–277. doi:10.1177/0956797611423545.

James, W. (1899). *Talks to teachers on psychology: And to students on some life's ideals.* New York, NY: Holt and Longmans.

Jennings, M. K., & Markus, G. B. (1977). The effect of military service on political attitudes: A panel study. *The American Political Science Review, 71*(1), 131–147. doi:10.2307/1956958.

Johansen, R. B., Laberg, J. C., & Martinussen, M. (2013). Measuring military identity: Scale development and psychometric evaluations. *Social Behavior and Personality: An International Journal, 41*(5), 861–880. doi:10.2224/sbp.2013.41.5.861.

John, O. P., Donahue, E., & Kentle, R. L. (1991). *The Big Five inventory: Versions 4a and 5a.* Technical Report. University of California, Berkeley, Institute of Personality and Social Research. Berkeley, CA.

Johnson, J. A. (1997). Units of analysis for the description and explanation of personality. In R. Hogan, J. Johnson, & S. Briggs (eds.), *Handbook of personality psychology* (pp. 73–93). San Diego, CA: Academic Press.

Johnston, C. S. (1995). The Rokeach value survey: Underlying structure and multidimensional scaling. *The Journal of Psychology, 129,* 583–597.

Jolliffe, I. T., & Cadima, J. (2016). Principal component analysis: A review and recent developments. *Philosophical Transactions A, 374*(2065), 20150202. doi:10.1098/rsta.2015.0202.

Kant, E. (1785). *Grundlegung zur Metaphysik der Sitten [Foundation of metaphysics of morality].* Ditzingen, Germany: Reclam.

Karp, D. R. (2000). Values theory and research. In E. F. Borgatta & R. J. V. Montgomery (eds.), *Encyclopedia of sociology* (pp. 3212–3227). New York, NY: MacMillan.

Kernic, F., & Annen, H. (2016). Führung und Werte [Leadership and values]. *Military Power Revue, 1,* 5–14.

King, L. A. (1995). Wishes, motives, goals, and personal memories: Relations of measures of human motivation. *Journal of Personality, 63,* 985–1007.

Kluckhohn, C. (1951). Values and Value-Orientations in the Theory of Action: An Exploration in Definition and Classification. In: Parsons, T. and Shils, E. (Eds.), *Toward a General Theory of Action*, Harvard University Press, Cambridge, 388–433. http://dx.doi.org/10.4159/harvard.9780674863507.c8

Kopelman, R. E., Rovenpor, J. L., & Guan, M. (2003). The study of values: Construction of the fourth edition. *Journal of Vocational Behavior*, 62(2), 203–220. doi:10.1016/S0001-8791(02)00047-7.

Kornguth, S., Steinberg, R., & Matthews, M. D. (2010). *Neurocognitive and physiological factors during high-tempo operations*. Aldershot, UK: Ashgate.

Krobath, H. T. (2009). *Werte [Values]*. Würzburg, Germany: Königshausen & Neumann.

Lam, S., Schaubroeck, J., & Aryee, S. (2002). Relationship between organizational justice and employee work outcomes: a cross-national study. *Journal of Organizational Behavior*, 23(1), 1–18. doi:10.1002/job.131.

Lang, F. R., Lüdtke, O., & Asendorpf, J. B. (2001). Testgüte und psychometrische Äquivalenz der deutschen Version des Big Five Inventory (BFI) bei jungen, mittelalten und alten Erwachsenen [Test quality and psychometric equivalence of the German version of the Big Five Inventory (BFI) among young, middle-aged and aged adults]. *Diagnostica*, 47(3), 111–121.

Lang, K. (1965). Military organizations. In J. G. March (Ed.), *Handbook of organizations* (pp. 838–878). Chicago, IL: Rand McNally.

Lee, M.-K. (2005). *Epistemology after protagoras: Responses to relativism in Plato, Aristotle, and Democritus*. New York, NY: Oxford University Press.

Lepine, J. A., Erez, A., & Johnson, D. E. (2002). The nature and dimensionality of organizational citizenship behavior: A critical review and meta-analysis. *Journal of Applied Psychology*, 87(1), 52–65. doi:10.1037//0021-9010.87.1.52.

Lickona, T. (1991). *Education for character*. New York, NY: Bantam.

Littman-Ovadia, H., & Lavy, S. (2012). Character strengths in Israel: Hebrew adaptation of the VIA inventory of strengths. *European Journal of Psychological Assessment*, 28(1), 41–50. doi:10.1027/1015-5759/a000089.

Liu, Y., & Cohen, A. (2010). Values, commitment, and OCB among Chinese employees. *International Journal of Intercultural Relations*, 34(5), 493–506. doi:10.1016/j.ijintrel.2010.05.001.

Locke, E. A., & Latham, G. P. (1990). Work motivation and satisfaction: Light at the end of the tunnel. *Psychological Science*, 1(4), 240–246. doi:10.1111/j.1467-9280.1990.tb00207.x.

Lurie, W. A. (1937). A study of Spranger's value-types by the method of factor analysis. *Journal of Social Psychology*, 8, 17–37.

Macdonald, C., Bore, M., & Munro, D. (2008). Values in action scale and the Big 5: An empirical indication of structure. *Journal of Research in Personality, 42*(4), 787–799. doi:10.1016/j.jrp.2007.10.003.

MacIntyre, A. C. (1981). *After virtue: A study in moral theory*. London, UK: Duckworth.

Maio, G. R., & Olson, J. M. (1998). Values as truisms: Evidence and implications. *Journal of Personality and Social Psychology, 74*, 294–311.

Manz, C. C., Anand, V., Joshi, M., & Manz, K. P. (2008). Emerging paradoxes in executive leadership: A theoretical interpretation of the tensions between corruption and virtuous values. *Leadership Quarterly, 19*(3), 385–392. doi:10.1016/j.leaqua.2008.03.009.

Matthews, M. D. (2008). Toward a positive military psychology. *Military Psychology, 20*, 289–298. doi:10.1080/08995600802345246.

Matthews, M. D. (2009). The soldier's mind: Motivation, mindset, and attidude. In A. R. Freeman, S. M. Freeman, & B. A. Moore (eds.), *Living and surviving in harm's way: A psychological treatment handbook for pre-and post deployment of military personnel* (pp. 9–26). New York, NY: Routledge.

Matthews, M. D. (2012). Cognitive and non-cognitive factors in soldier performance. In J. H. Laurence & M. D. Matthews (eds.), *The Oxford handbook of military psychology* (pp. 197–217). New York, NY: Oxford University Press.

Matthews, M. D. (2014). *Head strong: How psychology is revolutionizing war*. New York, NY: Oxford University Press.

Matthews, M. D., Brazil, D., & Erwin, M. S. (2009). *Character strengths and responding to leader challenges in combat*. Paper presented at the 21st Annual Convention of the Association for Psychological Science, San Francisco, CA.

Matthews, M. D., Eid, J., Kelly, D. R., Bailey, J. K., & Peterson, C. M. (2006b). Character strengths and virtues of developing military leaders: An international comparison. *Military Psychology, 18*(57), doi:10.1207/s15327876mp1803s_5.

Matthews, M. D., Lerner, R. M., & Annen, H. (2019). Non-cognitive amplifiers of human performance: Unpacking the 25/75 rule. In M. D. Matthews & D. M. Schnyer (eds.), *Human performance optimization: The science and ethics of enhancing human capabilities*. New York, NY: Oxford University Press.

Matthews, M. D., Peterson, C. M., & Kelly, D. R. (2006a). *Character strengths predicts retention of West Point cadets*. Paper presented at the American Society Meeting, New York, NY.

McAllister, I. (1995). Schools, enlistment, and military values: The Australian Services Cadet scheme. *Armed forces & society, 22*, 83–102.

McAllister, I., & Smith, H. (1989). Selecting the Guardians: Recruitment and military values. *Armed forces & society, 17*, 27–42.

McClelland, D. C. (1985). How motives, skills, and values determine what people do. *American Psychologist, 40*, 812–825.

McCrae, R. R., & Costa, P. T. (1990). *Personality in adulthood.* New York, NY: Guilford Press.

McCullough, M. E., & Snyder, C. R. (2000). Classical sources of human strength: Revisiting an old home and building a new one. *Journal of Social and Clinical Psychology, 19*(1), 1–10. doi:10.1521/jscp.2000.19.1.1.

McGrath, R. E. (2015). Integrating psychological and cultural perspectives on virtue: The hierarchical structure of character strengths. *The Journal of Positive Psychology, 10*(5), 407–424. doi:10.1080/17439760.2014.994222.

McGrath, R. E., & Walker, D. I. (2016). Factor structure of character strengths in youth: Consistency across ages and measures. *Journal of Moral Education, 7240*, 1–18. doi:10.1080/03057240.2016.1213709.

McGrath, R. E., Greenberg, M. J., & Hall-Simmonds, A. (2017). Scarecrow, Tin Woodsman, and Cowardly Lion: The three-factor model of virtue. *The Journal of Positive Psychology, 9760*, 1–20. doi:10.1080/17439760.2017.1326518.

McNeely, B. L., & Meglino, B. M. (1994). The role of dispositional and situational antecedents in prosocial organizational behavior: An examination of the intended beneficiaries of prosocial behavior. *Journal of Applied Psychology, 79*(6), 836–844. doi:10.1037/0021-9010.79.6.836.

Meglino, B. M., & Ravlin, E. C. (1998). Individual values in organizations: Concepts, controversies, and research. *Journal of Management, 24*(3), 351–389. doi:10.1016/S0149-2063(99)80065-8.

Meglino, B. M., Ravlin, E. C., & Adkins, C. L. (1991). Value congruence and satisfaction with a leader: An examination of the role of interaction. *Human Relations, 44*(5), 481–495. doi:10.1177/001872679104400504.

Meierhans, D., Rietmann, B., & Jonas, K. (2008). Influence of fair and supportive leadership behavior on commitment and organizational citizenship behavior. *Swiss Journal of Psychology, 67*, 131–141. doi:10.1024/1421-0185.67.3.131.

Meyer, E. G. (2015). The importance of understanding military culture. *Academic Psychiatry, 39*(4), 416–418. doi:10.1007/s40596-015-0285-1.

Mintz, S. M. (1996). Aristotelian virtue and business ethics education. *Journal of Business Ethics, 15*(8), 827–838. doi:10.1007/BF00381851.

Mohler, P. P., & Wohn, K. (2005). *Persönliche Werteorientierung im European social survey [Personal value orientation in the European social survey].* Mannheim, Germany: ZUMA.

Moorman, R. H., & Blakely, G. L. (1995). Individualism-Collectivism as an individual difference predictor of organizational citizenship behavior. *Journal of Organizational Behavior, 16*(16), 127–142.

Morales-Vives, F., De Raad, B., & Vigil-Colet, A. (2012). Psycholexical value factors in Spain and their relation with personality traits. *European Journal of Personality, 26,* 551–565. doi:10.1002/per.854.

Morales-Vives, F., De Raad, B., & Vigil-Colet, A. (2014). Psycho-lexically based virtue factors in Spain and their relation with personality traits. *The Journal of General Psychology, 141*(4), 297–325. doi:10.1080/00221309.2014.938719.

Morris, C. (1956). *Varieties of human value.* Chicago, IL: University of Chicago Press.

Morris, C., & Jones, L. V. (1955). Value scales and dimensions. *Journal of Abnormal and Social Psychology, 51,* 523–435.

Moskos, C. (1973). The emergent military: Civil, traditional, or plural? *The Pacific Sociological Review, 16*(2), 255–280. doi:10.1525/sop.2008.51.1.3.This.

Münsterberg, H. (1908). *Philosophie der Werte [Philosophy of values].* Leipzig, Germany: Johann Ambrosius Barth.

Neisser, U., Boodoo, G., Bouchard Jr, T. J., Boykin, A. W., Brody, N., Ceci, S. J., & Urbina, S. (1996). Intelligence: Knowns and unknowns. *American Psychologist, 51,* 77–101. doi: 10.1037/0003-066X.51.2.77.

Ng, V., Cao, M., Marsh, H. W., Tay, L., & Seligman, M. E. P. (2016). The factor structure of the Values in Action Inventory of Strengths (VIA-IS): An item-level exploratory structural equation modeling (ESEM) bifactor analysis. *Psychological Assessment, 29,* 1053–1058. doi:10.1037/pas0000396.

Nybert, D. (2007). The morality of everyday activities: Not the right, but the good thing to do. *Journal of Business Ethics, 81,* 587–598. doi:10.1007/s10551-007-9530-1.

Olsthoorn, P. (2011). *Military ethics and virtues. An interdisciplinary approach for the 21st century.* Oxon, UK: Routledge.

Olsthoorn, P. (2013). Virtue ethics in the military. In S. Van Hooft & N. Saunders (eds.), *The handbook of virtue ethics* (pp. 365–374). New York, NY: Routledge.

Omoto, A. M., & Snyder, M. (1995). Sustained helping without obligation: Motivation, longevity of service, and perceived attitude change among AIDS volunteers, Journal of Personality and Social Psychology, 68(4), 671–686.

O'Neil, D. P. (2007). *Predicting leader effectiveness: Personality and character strengths.* Unpublished doctoral dissertation, Durham, NC: Duke University.

O'Reilly, C., & Chatman, J. (1986). Organizational commitment and psychological attachment: The effects of compliance, identification, and internalization on prosocial behavior. *Journal of Applied Psychology, 71*(3), 492–499. doi:10.1037/0021-9010.71.3.492.

Organ, D. W. (1988). *Organizational citizenship behavior: The good soldier syndrome*. Lexington, MA: Lexington Books.

Organ, D. W. (1997). Organizational citizenship behavior: It's construct cleanup time. Human performance. *Human Performance, 10*, 85–97. doi:10.1207/s15327043hup1002_2.

Organ, D. W., Podsakoff, P. M., & Mackenzie, S. B. (2006). *Organizational citizenship behavior: Its nature, antecedents, and consequences*. Thousand Oaks, CA: Sage Publications.

Oyserman, D., Coon, H. M., & Kemmelmeier, M. (2002). Rethinking individualism and collectivism: Evaluation of theoretical assumptions and meta-analyses. *Psychological Bulletin, 128*, 3–72. doi:10.1037/0033-2909.128.1.3.

Park, N., & Peterson, C. M. (2007). Methodological issues in positive psychology and the assessment of character strengths. In A. D. Ong & M. H. M. Van Dulmen (eds.), *Handbook of methods in positive psychology* (pp. 292–305). New York, NY: Oxford University Press.

Park, N., & Peterson, C. M. (2009). Character strengths: Research and practice. *Journal of College and Character, 10*(8), 1–10. doi:10.2202/1940-1639.1042.

Park, N., & Peterson, C. M. (2010). Does it matter where we live?: The urban psychology of character strengths. *American Psychologist, 65*, 535–547. doi:10.1037/a0019621.

Park, N., Peterson, C. M., & Seligman, M. E. P. (2004). Strengths of character and well-being. *Journal of Social and Clinical Psychology, 23*(5), 603–619. doi:10.1521/jscp.23.5.603.50748.

Park, N., Peterson, C. M., & Seligman, M. E. P. (2006). Character strengths in fifty-four nations and the fifty US states. *Journal of Positive Psychology, 3*, 118–129. doi:10.1080/17439760600619567.

Parks, L., & Guay, R. P. (2009). Personality, values, and motivation. *Personality and Individual Differences, 47*, 675–684. doi.org/10.1016/j.paid.2009.06.002

Pathak, V., Rani, A., & Goswami, S. (2016). Value-based leadership. In N. Maheshwari & V. V. Kuma (eds.), *Military psychology concepts trends and interventions* (pp. 267–282). New Delhi, India: Sage Publications.

Pawelski, J. O. (2003). The promise of positive psychology for the assessment of character. *Journal of College and Character, 4*(3), doi:10.2202/1940-1639.1361.

Peabody, D. & Goldberg, L. (1989). Some determinants of factor structures from personality-trait descriptors. *Journal of personality and social psychology*. 57. 552–67. 10.1037//0022-3514.57.3.552.

Peterson, C. M. (2003). *The Values-in-Action Structured Interview (VIA-SI)*. Cincinnati, OH: The VIA Institute on Character.

Peterson, C. M. (2006). Character strengths. In C. M. Peterson (Ed.), *A primer in positive psychology* (pp. 137–164). Oxford, UK: Oxford University Press.

Peterson, C. M., & Park, N. (2003). Positive psychology as the evenhanded positive psychologist views it. *Psychological Inquiry, 14*, 143–147.

Peterson, C. M., Park, N., Pole, N., D'Andrea, W., & Seligman, M. E. P. (2008). Strengths of character and posttraumatic growth. *Journal of Traumatic Stress, 21*(2), 214–217. doi:10.1002/jts.20332.

Peterson, C. M., Park, N., & Seligman, M. E. P. (2005). Assessment of character strengths. In G. P. Koocher, J. C. Norcross, & S. S. Hill (eds.), *Psychologists' desk reference* (2nd ed., pp. 93–98). New York, NY: Oxford University Press.

Peterson, C. M., Ruch, W., Beermann, U., Park, N., & Seligman, M. E. P. (2007). Strengths of character, orientations to happiness, and life satisfaction. *The Journal of Positive Psychology, 2*, 149–156. doi:10.1080/17439760701228938.

Peterson, C. M., & Seligman, M. E. P. (2003). Character strengths before and after September 11. *Psychological Science, 14*, 381–384. doi:10.1111/1467-9280.24482.

Peterson, C. M., & Seligman, M. E. P. (2004). *Character strengths and virtues: A handbook and classification*. Oxford, UK: Oxford University Press.

Picano, J. & Roland, R. R. (2012). Assessing psychological suitability for high-risk military jobs. In J. H. Laurence & M. D. Matthews (eds.), *The Oxford handbook of military psychology* (pp. 148–157). New York, NY: Oxford University Press. doi:10.1093/oxfordhb/9780195399325.013.0056.

Picano, J., Roland, R. R., Rollins, K. D., & Williams, T. J. (2002). Development and validation of a sentence completion test measure of defensive responding in military personnel assessed for non-routine missions. *Military Psychology, 14*, 279–298.

Picano, J.J., Williams, T.J. & Roland, R.R. (2006). Assessment and selection of high-risk operational personnel. In C. H. Kennedy & E. A. Zillmer (eds.), *Military Psychology (pp.* 353–370). New York, NY: The Guilford Press.

Plato. (1968). *Republic* (translated by A. Bloom). New York, NY: Basic Books.

Podsakoff, N. P., Whiting, S. W., Podsakoff, P. M., & Blume, B. D. (2009). Individual- and organizational-level consequences of organizational citizenship behaviors: A meta-analysis. *Journal of Applied Psycholychol, 94*(1), 122–141. doi:10.1037/a0013079.

Podsakoff, P. M., MacKenzie, S. B., Lee, J.-Y., & Podsakoff, N. P. (2003). Common method biases in behavioral research: A critical review of the literature and recommended remedies, Journal of Applied Psychology, 88(5), 879–903. doi:10.1037/0021-9010.99.5.879.

Podsakoff, P. M., Mackenzie, S. B., Paine, J. B., & Bachrach, D. G. (2000). Organizational citizenship behaviors: A critical review of the theoretical and future research, Journal of Management, 26(3), 513–563. doi:10.1016/S0149-2063(00)00047-7.

Pols, H., & Oak, S. (2007). War & military mental health: The US psychiatric response in the 20th Century. American Journal of Public Health, 97, 2132–2142. doi:10.2105/AJPH.2006.090910.

Pozzebon, J. A., & Ashton, M. C. (2009). Personality and values as predictors of self- and peer-reported behavior. Journal of Individual Differences, 30, 122–129. doi:10.1027/1614-0001.30.3.122.

Proyer, R. T., Annen, H., Eggimann, N., Schneider, A., & Ruch, W. (2012). Assessing the "Good Life" in a military context: How does life and work-satisfaction relate to orientations to happiness and career-success among Swiss professional officers? Social Indicators Research, 106, 577–590. doi:10.1007/s11205-011-9823-8.

Rammstedt, B. (2007). Who worries and who is happy? Explaining individual differences in worries and satisfaction by personality. Personality and Individual Differences, 43, 1626–1634. doi:10.1016/j.paid.2007.04.031.

Ray, J. R. (1973). Conservatism, authoritarianism, and related variables: A review and empirical study. In G. Wilson (Ed.), The psychology of conservatism (pp. 17–33). London, UK: Routledge.

Renner, W. (2003a). A German value questionnaire developed on a lexical basis: Construction and steps toward a validation. Review of Psychology, 10, 107–123.

Renner, W. (2003b). Human values: A lexical perspective. Personality and Individual Differences, 34(1), 127–141. doi:10.1016/S0191-8869(02)00037-5.

Renner, W., & Myambo, K. (2007). The Arabic language and contemporary Egyptian national values: A lexical analysis. Psychologia, 50(1), 26–38. doi:10.2117/psysoc.2007.26.

Renner, W., Peltzer, K., & Phaswana, M. G. (2003). The structure of values among Northern Sotho speaking people in South Africa. South African Journal of Psychology, 33(2), 103–108. doi:10.1177/008124630303300205.

Renner, W., Salem, I., & Alexandrowicz, R. (2004). Human values as predictors for political, religious and health-related attitudes: A contribution towards validating the Austrian Value Questionnaire (AVQ) by structural equation modeling. *Social Behavior and Personality, 32*(5), 477–490. doi:10.2224/sbp.2004.32.5.477.

Rice, C. (2014). *Speech by Condolezza Rice on the occasion of the awarding of the Thayer Award, U.S. Military Academy, West Point (2014, 6 October).* Retrieved December 20, 2017, from https://www.westpointaog.org/about/awards/thayerawardcondoleezzarice2014speech.

Richards, J. M. (1966). Life goals of American college freshmen. *Journal of Counseling Psychology, 13*, 12–20.

Rioux, S. M., & Penner, L. A. (2001). The causes of organizational citizenship behavior: A motivational analysis. *Journal of Applied Psychology, 86*(6), 1306–1314. doi:10.1037/0021-9010.86.6.1306.

Roberts, B. W., & Robins, R. W. (2001). Broad dispositions, broad aspirations: The intersection of personality traits and major life goals. *Personality and Social Psychology Bulletin, 26*, 1284–1296.

Robinson, P. (2008). Introduction: Ethics education in the military. In P. Robinson, N. de Lee, & D. Carrick (eds.), *Ethics education in the military* (pp. 11–12). New York, NY: Routledge.

Roccas, S., Sagiv, L., Schwartz, S. H., & Knafo, A. (2002). The Big Five personality factors and personal values. *Personality and Social Psychology Bulletin, 28*(6), 789–801. doi:10.1177/0146167202289008.

Rohan, M. J. (2000). A rose by any name? The values construct. *Personality and Social Psychology Review, 4*(3), 255–277. doi:10.1207/S15327957PSPR0403_4.

Rokeach, M. (1967). *Value survey*. Sunnyvale, CA: Halgren Tests.

Rokeach, M. (1973a). *Rokeach value survey*. San Francisco, CA: Health. doi:10.1037/t01381-000.

Rokeach, M. (1973b). *The nature of human values*. New York, NY: The Free Press.

Rokeach, M. (1974). Change and stability of American value systems, 1968–71. *Public Opinion Quarterly, 38*, 232–238.

Rokeach, M. (1979). *Understanding human values: Individual and societal*. New York, NY: The Free Press.

Rokeach, M., & Ball-Rokeach, S. J. (1989). Stability and change in American value priorities 1968–81. *American Psychologist, 44*, 775–784.

Ross, P. T., Ravindranath, D., Clay, M., & Lypson, M. L. (2015). A greater mission: Understanding military culture as a tool for serving those who have served. *Journal of Graduate Medical Education, 7*(4), 519–522. doi:10.4300/JGME-D-14-00568.1.

Ruch, W., & Proyer, R. T. (2015). Mapping strengths into virtues: The relation of the 24 VIA-strengths to six ubiquitous virtues. *Frontiers in Psychology, 6*(3), 1–12. doi:10.3389/fpsyg.2015.00460.

Ruch, W., Proyer, R. T., Harzer, C., Park, N., Peterson, C. M., & Seligman, M. E. P. (2010). Values in Action Inventory of Strengths (VIA-IS). *Journal of Individual Differences, 31*(3), 138–149. doi:10.1027/1614-0001/a000022.

Ruch, W., Weber, M., Park, N., & Peterson, C. M. (2014). Character strengths in children and adolescents: Reliability and initial validity of the German Values in Action Inventory of Strengths for Youth (German VIA-Youth). *European Journal of Psychological Assessment, 30,* 57–64. doi: 10.1027/1015-5759/a000169.

Russi, F. (2009). *Über Werte und Tugenden [On values and virtues].* Weimar, Germany: Bertuch Verlag.

Salem, I., & Renner, W. (2004). Do human values reflect job decisions and prosocial and antisocial behavior? A contribution towards validating the Austrian Value Questionnaire by group comparisons. *Psychol Rep, 94*(3), 995–1008. doi:10.2466/pr0.94.3.995-1008.

Sandage, S. J., & Hill, P. C. (2001). The virtues of positive psychology: The rapprochement and challenges of an affirmative postmodern perspective. *Journal for the Theory of Social Behaviour, 31*(3), 241–260. doi:10.1111/1468-5914.00157.

Sandin, P. (2007). Collective military virtues. *Journal of Military Ethics, 6*(4), 303–314. doi:10.1080/15027570701755505.

Saucier, G. (2000). Isms and the structure of social attitudes. *Journal of Personality and Social Psychology, 78*(2), 366–385. doi:10.1037//0022-3514.78.2.366

Saucier, G., Hampson, S. E., & Goldberg, L. R. (2000). Cross-language studies of lexical personality factors. In S. E. Hampson (Ed.), *Advances in personality psychology* (Vol. 1, pp. 1–36). Philadelphia; PA: Taylor & Francis.

Saucier, G., & Srivastava, S. (2015). What makes a good structural model of personality? Evaluating the Big Five and alternatives. In M. Mikulincer & P. R. Shaver (eds.), *APA handbook of personality and pocial psychology: Vol. 4. personality processes and individual differences* (pp. 283–305). doi:10.1037/14343-013.

Saucier, G., Thalmayer, A. G., Payne, D. L., Carlson, R., Sanogo, L., Ole-Kotikash, L., Zhou, L. (2014). A basic bivariate structure of personality attributes evident across nine widely diverse languages. *Journal of Personality, 82,* 1–14. doi:10.1111/jopy.12028.

Scales, R. H. (2009). Clausewitz and World War IV. *Military Psychology, 21*(Suppl 1), 23–35. doi:10.1080/08995600802554573.

Schein, E. H. (1985). *Organizational culture and leadership: A dynamic view.* San Francisco, CA: Jossey-Bass.

Schein, E. H. (1992). *Organizational Culture and Leadership,* San Francisco, CA: Jossey Bass.

Schlöder, B. (1993). *Soziale Werte und Werthaltungen [Social values and value orientation].* Berlin, Germany: Springer.

Schmidt, P., Bamberg, S., Davidov, E., Herrmann, J., & Schwartz, S. H. (2007). Die Messung von Werten mit dem "Portraits Value Questionnaire" [The measurement of values with the "Portraits Value Questionnaire"]. *Zeitschrift für Sozialpsychologie, 38*(4), 261–275. doi:10.1024/0044-3514.38.4.261.

Scholl-Schaaf, M. (1975). *Werthaltung und Wertsystem: Ein Plädoyer für die Verwendung des Wertbegriffs in der Sozialpsychologie [A plea for the use of the value concept in social psychology].* Bonn, Germany: Bouvier.

Schumm, W. R., Gade, P. A., & Bell, D. B. (2003). Dimensionality of military professional values items: An exploratory factor analysis of data from the spring 1996 Sample Survey of Military Personnel. *Psychological Reports, 92,* 831–841. doi:10.2466/pr0.2003.92.3.831.

Schwartz, S. H. (1992). Universals in the content and structure of values: Theoretical advances and empirical tests in 20 countries. In M. P. Zanna (Ed.), *Advances in experimental social psychology 25* (pp. 1–65). New York, NY: Academic Press.

Schwartz, S. H. (1994). Are there universal aspects in the structure and content of human values? *Journal of Social Issues, 50,* 19–45. doi:10.1111/j.1540-4560.1994.tb01196.x.

Schwartz, S. H. (1999). A theory of cultural values and some implications for work. *Applied Psychology, 48,* 23–47. doi:10.1111/j.1464-0597.1999.tb00047.x.

Schwartz, S. H. (2005). Robustness and fruitfulness of a theory of universals in individual values. *Valores E Trabalho, 921,* 56–85.

Schwartz, S. H. (2006). A theory of cultural value orientations: Explication and applications. *Comparative Sociology, 5,* 2(3), 137–182. doi:10.1163/156913306778667357.

Schwartz, S. H. (2012). An overview of the Schwartz theory of basic values. *Online Readings in Psychology and Culture, 2,* 1–20. doi:10.9707/2307-0919.1116.

Schwartz, S. H., & Bilsky, W. (1987). Toward a universal psychological structure of human values. *Journal of Personality and Social Psychology, 53*(3), 550–562. doi:10.1037/0022-3514.53.3.550.

Schwartz, S. H., & Bilsky, W. (1990). Toward a theory of the universal content and structure of values: Extensions and cross-cultural replications. *Journal of Personality and Social Psychology, 58*(5), 878–891. doi:10.1037/0022-3514.58.5.878.

Schwartz, S. H., Sagiv, L., & Boehnke, K. (2000). Worries and values. *Journal of Personality, 68*(2), 309–346. doi:10.1111/1467-6494.00099.

Seligman, M. E. P. (1998). Building human strength: Psychology's forgotten mission. *American Psychological Association, 2.* doi:10.1037/10288-000.

Seligman, M. E. P. (2000). Positive psychology. In J. E. Gillham (Ed.), *The science of optimism and hope: Research essays in honor of Martin E. P. Seligman* (pp. 415–429). Radnor, PA: Templeton Foundation Press.

Seligman, M. E. P. (2002). Positive psychology, positive prevention, and positive therapy. In S. J. Lopez & C. R. Snyder (eds.), *Handbook of positive psychology* (pp. 3–12). New York, NY: Oxford University Press.

Seligman, M. E. P., & Csikszentmihalyi, M. (2000). Positive psychology: An introduction. *American Psychologist, 55,* 5–14. doi:10.1037/0003-066X.55.1.5.

Seligman, M. E. P., Steen, T. A., Park, N., & Peterson, C. M. (2005). Positive psychology progress: Empirical validation of interventions. *American Psychologist, 60*(5), 410–421. doi:10.1037/0003-066X.60.5.410.

Settersten, R. A., Jr. (2006). When nations call: How wartime military service matters for the life course and aging. *Research on Aging, 28*(1), 12–36.

Shell. (2015). *Shell Youth Study 2015.* Retrieved September 29, 2018, from https://www.shell.de/ueber-uns/die-shell-jugendstudie.html.

Shryack, J., Steger, M. F., Krueger, R. F., & Kallie, C. S. (2010). The structure of virtue: An empirical investigation of the dimensionality of the virtues in action inventory of strengths. *Personality and Individual Differences, 48*(6), 714–719. doi:10.1016/j.paid.2010.01.007.

Smolicz, J. (1981). Core values and cultural identity. *Ethnic and Racial Studies, 4*(1), 75–90. doi:10.1080/01419870.1981.9993325.

Snider, D., & Matthews, L. J. (2012). *The future of the army profession.* 2nd ed. Boston, MA: McGraw-Hill.

Soeters, J. (1997). Values in military academies. A thirteen country study. *Armed Forces & Society, 24,* 7–32.

Soeters, J. L., Poponete, C., & Page, J. (2006). Culture's consequences in the military. In T. W. Britt, A. B. Adler, & C. A. Castro (eds.), *Military life. The psychology of serving in peace and combat* (Vol. 4, pp. 13–34). Westport, CT: Praeger Security.

Soeters, J. L., Winslow, D. J., & Weibull, A. (2006). Military culture. In G. Caforio (Ed.), *Handbook of the sociology of the military* (pp. 237–254). New York, NY: Springer US.

Solomon, R. C. (1992). Corporate roles, personal virtues: An Aristotelean approach to business ethics. *Business Ethics Quarterly, 2*, 317–339.

Spranger, E. (1928). *Types of men.* New York, NY: Stechert-Hafner.

Staffelbach, B. (2006). Effizienz der (Kader-) Ausbildung in der (Miliz-) Armee [Efficiency in training of military cadre in the militia army]. In H. Annen & U. Zwygart (eds.), *Das Ruder in der Hand. Aspekte der Führung und Ausbildung in Armee, Wirtschaft und Politik* [*The oar in hand. Aspects of management and training in the military, economy, and politics*] (pp. 99–104). Frauenfeld, Switzerland: Huber.

Staub, E. (1989). Individual and societal (group) values in a motivational perspective and their role in benevolence and harmdoing. In , J. R. Eisenberg & B. Staub (eds.), *Social and moral values: Individual and societal perspectives* (pp. 45–61). Hillsdale, NJ: Erlbaum.

Steinberg, R., & Kornguth, S. E. (2009). Sustaining performance under stress: Overview of this issue. *Military Psychology, 21*(S1), doi:10.1080/08995600802554516.

Stocker, D., Jacobshagen, N., Semmer, N. K., & Annen, H. (2010). Appreciation at work in the Swiss Armed Forces. *Swiss Journal of Psychology, 69*, 117–124. doi:10.1024/1421-0185/a000013.

Strack, M., Gennerich, C., & Hopf, N. (2008). Warum Werte? [Why values?]. In E. H. Witte (Ed.) *Sozialpsychologie und Werte* [*Social psychology and values*] (pp. 90–130). Lengerich, Germany: Pabst.

Strelau, J., & Zawadzki, B. (1996). Temperament dimensions as related to the giant three and the Big Five factors: A psychometric approach. *Life and Scientific Creativity, 3*, 260–281.

Swiss Armed Forces. (1994). *Dienstreglement 04 der Schweizer Armee* [*Service regulations of the Swiss Armed Forces of 22 June 1994*]. Bern, Switzerland: Government Printing Office.

Swiss Armed Forces. (2010). *Militärethik in der Schweizer Armee* [*Military ethics in the Swiss Armed Forces*]. *Bericht des Bundesrates über die innere Führung der Armee in Erfüllung des Postulates 05.3060 Widmer vom 10. März 2005* [*Report of the Federal Council on the internal management*]. Bern, Switzerland: Government Printing Office.

Swiss Armed Forces. (2012). Fragebogen für Führungsmotivation [questionnaire for motivation to lead], In Swiss Armed Forces (ed.), *Qualifikations-und Mutationswesen in der Armee [Qualifications and redeployments redeployments in the Swiss Armed Forces]*. Berne, Switzerland: BBL.

Swiss Confederation. (1999). *Bundesverfassung der Schweizerischen Eidgenossenschaft [Federal constitution of the Swiss confederation]*. Bern, Switzerland: Government Printing Office.

Szvircsev Tresch, T. (2011). The transformation of Switzerland's militia armed forces and the role of the citizen in uniform. *Armed Forces & Society, 37*, 239–260. doi:10.1177/0095327x10361670.

Thomas, J. L., Dickson, M. W., & Bliese, P. D. (2001). Values predicting leader performance in the U.S. Army reserve officer training corps assessment center: Evidence for a personality-mediated model. *The Leadership Quarterly, 12*, 181–196.

Tjeltveit, A. C. (2003). Implicit virtues, divergent goods, multiple communities: Explicitly addressing virtues in the behavioral sciences. *American Behavioral Scientist, 47*(4), 395–414. doi:10.1177/0002764203256946.

Trommsdorff, G. (1996). Werte und Wertewandel im kulturellen Kontext aus psychologischer Sicht [Values and value shift in the cultural context from a psychological perspective]. In E. Janssen (Ed.), *Gesellschaft im Umbruch? Aspekte des Wertewandels in Deutschland, Japan und Osteuropa [Society in transition? Aspects of value shift in in Germany, Japan and Eastern Europe]* (pp. 13–40). München, Germany: Iudicium.

Urban, W. M. (1907). Recent tendencies in the psychological theory of values. *Psychological Bulletin, 4*(3), 65–72.

U.S. Department of the Army. (1999). *Military leadership* (Field Manual 22-100). Washington, D.C.: Headquarters, Department of the Army.

U.S. Military Academy. (n.d.). Cadet Prayer. Retrieved December 20, 2017, from https://www.usma.edu/chaplain/sitepages/cadet%20prayer.aspx.

Van Deth, J. W. (1998). Introduction: The impact of values. *The Impact of Values*, 1–18. doi:10.1093/0198294751.003.0001.

Van Dyne, L., & LePine, J. (1998). Predicting voice behavior in work groups. *Journal of Applied Psychology, 83*(6), 853.

Van Eeden, C., Wissing, M. P., Dreyer, J., Park, N., & Peterson, C. M. (2008). Validation of the Values in Action Inventory of Strengths for Youth (VIA-Youth) among South African learners. *Journal of Psychology in Africa, 18*(1), 143–154. doi:10.1080/14330237.2008.10820181.

Van Eijnatten, F. M., Van der Ark, L. A., Holloway, S. S., Van Eijnatten, F. M., & Ark, L. (2015). Ipsative measurement and the analysis of organizational values: An alternative approach for data analysis. *Quality & Quantity*, 49(2), 1–21. doi:10.1007/s11135-014-0009-8.

Van Iddekinge, C. H., Ferris, G. R., & Hefner, T. S. (2009). Test of a multi-stage model of distal and proximal predictors of leader performance. *Personnel Psychology*, 62, 461–493.

Van Oudenhoven, J. P., De Raad, B., Carmona, C., Helbig, A. K., & Van der Linden, M. (2012). Are virtues shaped by national cultures or religions? *Swiss Journal of Psychology*, 71(1), 29–34. doi:10.1024/1421-0185/a000068.

Velicer, W. F. (1976). Determining the number of components from the matrix of partial correlations. *Psychometrika*, 41, 321–327.

Vernon, P. E., & Allport, G. W. (1931). A test for personal values. *The Journal of Abnormal and Social Psychology*, 26, 231–248. doi:10.1037/h0073233.

Vroom, V. H. (1964). *Work and motivation*. New York, NY: Wiley.

Walker, L. J., & Pitts, R. C. (1998). Naturalistic conceptions of moral maturity. *Developmental Psychology*, 34(3), 403–419. doi:10.1037/0012-1649.34.3.403.

Wiggins, J. S. (1991). Agency and communion as conceptual coordinates for the understanding and measurement of interpersonal behavior. In W. Grove & D. Cicchetti (eds.), *Thinking clearly about psychology: Essays in honor of Paul E. Meehl* (pp. 89–113). Minneapolis, MN: University of Minnesota Press.

Wiggins, J. S. (2003). *Paradigms of personality*. New York, NY: Guilford.

Wille, U. (1854). Erziehung und Wesen des Offiziers: Zum 70. Geburtstag unseres Generals am 5. April 1918 [Education and nature of an officer: *To the 70th birthday of our Brigadier General on April 5th 1918*]. Schweizerische Militärzeitschrift, 14, 105–106.

Williams, R. M. (1979). Change and stability in values and value systems: A sociological perspective. In M. Rokeach (Ed.), *Understanding human values* (pp. 15–46). New York, NY: Free Press.

Wirtz, M. A. (2016). Values. In M. A. Wirtz (ed.), *Dorsch* (p. 1790). Bern, Switzerland: Hogrefe.

Yearley, L. H. (1990). *Mencius and Aquinas: Theories of virtue and conceptions of courage*. Albany, NY: State University of New York Press.

Zagzebski, L. T. (1996). *Virtues of the mind: An inquiry into the nature of virtue and the ethical foundations of knowledge*. Cambridge, UK: Cambridge University Press.

Studies in Military Psychology and Pedagogy

Editor: Dr. H. Annen

Band 1   Edmund A. van Trotsenburg: Militärpädagogik. 1989.

Band 2   Hermann Jung / Heinz Florian: Grundlagen der Militärpädagogik. Eine Anleitung zu pädagogisch verantwortetem Handeln. 1994.

Band 3   Franz Kernic: Demokratie und Wehrsystem. Aufsätze zum Verhältnis von Gesellschaft, politischem System und Heer in Österreich. 1997.

Band 4   Edwin R. Micewski: Grenzen der Gewalt – Grenzen der Gewaltlosigkeit. Zur Begründung der Gewaltproblematik im Kontext philosophischer Ethik und politischer Philosophie. 1998.

Band 5   Rudolf Egger / Heinz Florian (Hrsg.): Pädagogische Professionalisierung im Bundesheer. Dokumentation und Reflexion des PädAk-Sonderstudienganges Wehrpädagogisches Management. 1999.

Band 6   Franz Kernic / Harald Haas: Warriors for Peace. A Sociological Study on the Austrian Experience of UN Peacekeeping. 1999.

Band 7   Franz Kernic / Jean M. Callaghan / Philippe Manigart: Public Opinion on European Security and Defense. A Survey of European Trends and Public Attitudes Toward CFSP and ESDP. 2002.

Band 8   Heinz Florian (ed.): Military Pedagogy – An International Survey. 2002.

Band 9   Edwin R. Micewski / Hubert Annen (eds.): Military Ethics in Professional Military Education – Revisited. 2005.

Band 10 Hubert Annen / Wolfgang Royl (eds.): Military Pedagogy in Progress. 2007.

Band 11 Hubert Annen / Wolfgang Royl (eds.): Educational Challenges Regarding Military Action. 2010.

Band 12 Hubert Annen / Juha Mäkinen / Can Nakkas (eds.): Thinking and Acting in Military Pedagogy. 2013.

Band 13 Duraid Jalili / Hubert Annen (eds.): Professional Military Education. A Cross-Cultural Survey. 2019.

Band 14 Nadine Eggimann Zanetti: Values and Virtues in the Military. 2020.

www.peterlang.com

www.ingramcontent.com/pod-product-compliance
Ingram Content Group UK Ltd.
Pitfield, Milton Keynes, MK11 3LW, UK
UKHW021829210426
5322IPUK00004B/98